STRENGTHENING HIGH SCHOOL CHEMISTRY EDUCATION THROUGH TEACHER OUTREACH PROGRAMS

A WORKSHOP SUMMARY TO THE CHEMICAL SCIENCES ROUNDTABLE

Steve Olson, Rapporteur

Chemical Sciences Roundtable

Board on Chemical Sciences and Technology

Division on Earth and Life Studies

D1451059

NATIONAL RESEARCH COUNCIL
OF THE NATIONAL ACADEMIES

THE NATIONAL ACADEMIES PRESS
Washington, D.C.
www.nap.edu

THE NATIONAL ACADEMIES PRESS   **500 Fifth Street, N.W.**   **Washington, DC 20001**

NOTICE: The project that is the subject of this report was approved by the Governing Board of the National Research Council, whose members are drawn from the councils of the National Academy of Sciences, the National Academy of Engineering, and the Institute of Medicine. The members of the committee responsible for the report were chosen for their special competences and with regard for appropriate balance.

This study was supported by the U.S. Department of Energy under Grant DE-FG02-07ER15872, the National Institutes of Health under Grant N01-OD-4-2139 (Task Order 25), and the National Science Foundation under Grant CHE-0621582.

Any opinions, findings, conclusions, or recommendations expressed in this publication are those of the authors and do not necessarily reflect the views of the organizations or agencies that provided support for the project.

International Standard Book Number-13:   978-0-309-12859-9
International Standard Book Number-10:   0-309-12859-5

Additional copies of this report are available from the National Academies Press, 500 Fifth Street, N.W., Lockbox 285, Washington, DC 20055; (800) 624-6242 or (202) 334-3313 (in the Washington metropolitan area); Internet, http://www.nap.edu.

THE NATIONAL ACADEMIES
Advisers to the Nation on Science, Engineering, and Medicine

The **National Academy of Sciences** is a private, nonprofit, self-perpetuating society of distinguished scholars engaged in scientific and engineering research, dedicated to the furtherance of science and technology and to their use for the general welfare. Upon the authority of the charter granted to it by the Congress in 1863, the Academy has a mandate that requires it to advise the federal government on scientific and technical matters. Dr. Ralph J. Cicerone is president of the National Academy of Sciences.

The **National Academy of Engineering** was established in 1964, under the charter of the National Academy of Sciences, as a parallel organization of outstanding engineers. It is autonomous in its administration and in the selection of its members, sharing with the National Academy of Sciences the responsibility for advising the federal government. The National Academy of Engineering also sponsors engineering programs aimed at meeting national needs, encourages education and research, and recognizes the superior achievements of engineers. Dr. Charles M. Vest is president of the National Academy of Engineering.

The **Institute of Medicine** was established in 1970 by the National Academy of Sciences to secure the services of eminent members of appropriate professions in the examination of policy matters pertaining to the health of the public. The Institute acts under the responsibility given to the National Academy of Sciences by its congressional charter to be an adviser to the federal government and, upon its own initiative, to identify issues of medical care, research, and education. Dr. Harvey V. Fineberg is president of the Institute of Medicine.

The **National Research Council** was organized by the National Academy of Sciences in 1916 to associate the broad community of science and technology with the Academy's purposes of furthering knowledge and advising the federal government. Functioning in accordance with general policies determined by the Academy, the Council has become the principal operating agency of both the National Academy of Sciences and the National Academy of Engineering in providing services to the government, the public, and the scientific and engineering communities. The Council is administered jointly by both Academies and the Institute of Medicine. Dr. Ralph J. Cicerone and Dr. Charles M. Vest are chair and vice chair, respectively, of the National Research Council.

www.national-academies.org

CHEMICAL SCIENCES ROUNDTABLE

Co-chairs

Charles P. Casey, University of Wisconsin, Madison
Sharon Haynie, E. I. du Pont de Nemours & Company, Wilmington, Delaware

Members

Patricia A. Baisden, Lawrence Livermore National Laboratory, Livermore, California
Mark A. Barteau, University of Delaware, Newark, Delaware
Michael R. Berman, Air Force Office of Scientific Research, Arlington, Virginia
Apurba Bhattacharya, Texas A&M, Kingsville, Texas
Louis Brus, Columbia, New York
Paul F. Bryan, Biofuels Technology Chevron Technology Ventures LLC, Richmond, California
Mark Cardillo,* Camille & Henry Dreyfus Foundation, New York
William F. Carroll Jr.,* Occidental Chemical Corporation, Dallas, Texas
Marvin H. Caruthers, University of Colorado, Boulder, Colorado
John C. Chen, Lehigh University, Bethlehem, Pennsylvania
Luis Echegoyen, National Science Foundation, Arlington, Virginia
Barbara J. Finlayson-Pitts, University of California, Irvine, California
Gary J. Foley, U. S. Environmental Protection Agency, Research Triangle Park, North Carolina
Teresa Fryberger, NASA Earth Sciences Division, Washington, District of Columbia
Alex Harris,* Brookhaven National Laboratory, Upton, New York
Luis E. Martinez, The Scripps Research Institute, Jupiter Florida
John J. McGrath, National Science Foundation, Arlington, Virginia
Paul F. McKenzie, Centocor R&D, Radnor, New Jersey
Douglas Ray, Pacific Northwest National Laboratory, Richland, Washington
Michael E. Rogers, National Institutes of Health, Bethesda, Maryland
Eric Rolfing, U.S. Department of Energy, Washington, District of Columbia
James M. Solyst, ENVIRON International Corporation, Arlington, Virginia
Levi Thompson, University of Michigan, Ann Arbor

National Research Council Staff

Dorothy Zolandz, Director
Andrew Crowther, Postdoctoral Fellow
Tina M. Masciangioli, Responsible Staff Officer
Jessica Pullen, Administrative Assistant
Sheena Siddiqui, Research Assistant
Lynelle Vidale, Program Assistant

---

* These members of the Chemical Sciences Roundtable oversaw the planning of the Workshop on Strengthening High School Chemistry Education Through Teacher Outreach Programs but were not involved in the writing of this workshop summary.

Preface

The Chemical Sciences Roundtable (CSR) was established in 1997 by the National Research Council. It provides a science-oriented apolitical forum for leaders in the chemical sciences to discuss chemistry-related issues affecting government, industry, and universities. Organized by the National Research Council's Board on Chemical Sciences and Technology, the CSR aims to strengthen the chemical sciences by fostering communication among the people and organizations—spanning industry, government, universities, and professional associations—involved with the chemical enterprise. One way it does this is by organizing workshops that address issues in chemical science and technology that require national attention.

In August 2008, the CSR organized a workshop on the topic, "Strengthening High School Chemistry Education through Teacher Outreach Program." The workshop brought together representatives of government, industry, academia, scientific societies, and foundations who are involved in organizing, funding, and delivering in-service outreach programs for high school chemistry teachers. The goal of the workshop was to explore how high school chemistry education could be improved through teacher outreach programs, with a particular emphasis on assessments of program effectiveness. The workshop sought programs that could improve the chemistry education of all students, not just those pursing a career in science. To this end, presentations at the workshop covered the current status of high school chemistry education; provided examples of public and private outreach programs; and explored how to evaluate whether current outreach programs are meeting the needs of chemistry teachers and students. The workshop did not attempt to address the many other issues related to high school chemistry education, including pre-service teacher training, national standards, teacher compensation, and teacher shortages.

This document summarizes the presentations and discussions that took place at the workshop, and includes poster presenter abstracts. In accordance with the policies of the CSR, the workshop *did not* attempt to establish any conclusions or recommendations about needs and future directions, focusing instead on issues identified by the speakers. In addition, the organizing committee's role was limited to planning the workshop. The workshop summary has been prepared by the workshop rapporteur Steve Olsen as a factual summary of what occurred at the workshop.

Acknowledgment of Reviewers

This report has been reviewed in draft form by persons chosen for their diverse perspectives and technical expertise in accordance with procedures approved by the National Research Council's Report Review Committee. The purpose of this independent review is to provide candid and critical comments that will assist the institution in making the published report as sound as possible and to ensure that it meets institutional standards for objectivity, evidence, and responsiveness to the study charge. The review comments and draft manuscript remain confidential to protect the integrity of the deliberative process. We wish to thank the following individuals for their review of this report:

Paul Bryan, Chevron Technology Ventures LLC, Richmond, California
John Chen, Lehigh University, Bethlehem, Pennsylvania
Eric Jakobsson, University of Illinois, Urbana
Steven Long, Rogers High School, Rogers, Arkansas

Although the reviewers listed above provided many constructive comments and suggestions, they were not asked to endorse the conclusions or recommendations nor did they see the final draft of the report before its release. The review of this report was overseen by **Elizabeth A. Carvellas**, Teacher Advisory Council. Appointed by the National Research Council, she was responsible for making certain that an independent examination of this report was carried out in accordance with institutional procedures and that all review comments were carefully considered. Responsibility for the final content of this report rests entirely with the authoring committee and the institution.

Contents

1

Overview

This day-and-a-half workshop began with an introduction by the workshop organizers **Mark Cardillo**, Dreyfus Foundation; **William (Bill) Carroll**, Occidental Chemical Corporation; and **Alex Harris**, Brookhaven National Laboratory. They emphasized the challenge of addressing such a broad and sweeping topic as high school chemistry education. This led them to focus on in-service teacher outreach programs for high school education, because high school is where chemistry becomes a discrete discipline and outreach programs are a potential conduit by which the greater chemical community can make a contribution. Particular emphasis was placed on evaluations of the effectiveness of these programs.

Finally, Harris outlined the workshop organization. Day 1 consisted of two sessions and a poster session, and day 2 included a third session. Session 1 addressed the question What are the major and general issues in high school chemistry education? It included remarks on the current state of science and the importance of teachers. Sessions 2 and 3 addressed the question Who is doing what with respect to high school chemistry education (and how is effectiveness measured)? Session 2 focused on publicly funded government agency and university programs. Session 3 addressed privately funded for-profit and nonprofit programs.

Session 1 began with an overview of the state of science and science education in the United States, provided by **Kathryn Sullivan,** Battelle Center for Mathematics and Science Education Policy. Sullivan presented the current position of the United States in research and development and in scientific and mathematics education. She showed that research and development (R&D) have become more internationally distributed even as R&D in the United States has grown substantially in scale and scope. The need for more professional development opportunities for teachers was discussed. She also talked about the role of student and parent attitudes in education. For example, parents recognize the need for improved science and mathematics education but tend to be satisfied with the amount of science and mathematics their own children study in school. The National Science Board has identified better coordination and more effective teaching as the greatest needs of the U.S. educational system.

Session 1 continued with a presentation by **Robert Tai,** University of Virginia, on the current state of high school chemistry education. **Gerry Wheeler,** National Science Teachers Association, and **Roxie Allen,** Associated Chemistry Teachers of Texas, provided the national and state-level teachers' perspective, respectively. The session concluded with a local teacher panel composed of **Caryn Galatis**, Thomas A. Edison High School, Virginia; **Brian J. Kennedy**, Thomas Jefferson High School for Science and Technology, Virginia; and **Kiara Hargrove**, Baltimore Polytechnic Institute, Maryland. Tai's presentation of longitudinal data demonstrated that exposure to particular subjects in high school chemistry, frequent peer interactions, and studying high-level mathematics are positively associated with chemistry grades in college, while time spent on community and student projects, labs, and instructional technologies can be negatively associated with college chemistry grades. He also showed that most high school chemistry teachers have taken college courses above the level they are assigned to teach, but they report needing help in using technology in science instruction, teaching classes with special needs students, and using inquiry-oriented teaching methods. Speakers indicated that laboratories in high school chemistry tend to be disconnected from coursework, focus on procedures rather than clear learning outcomes, and provide few opportunities for discussion or reflection. Across the country, new requirements that high school students take more advanced science courses have increased the need for well-prepared chemistry teachers. Teachers feel that a major challenge for high school chemistry teachers is connecting the subject to everyday experiences, and professional development that focuses on this linkage can be especially valuable.

The first half of Session 2 focused on publicly funded programs at government agencies. The presenters were **L. Anthony Beck**, National Institutes of Health (NIH); **Katherine Covert** and **Joan Prival**, National Science Foundation (NSF); **Jeffery Dilks**, Department of Energy (DOE); and **Kenneth White**, Brookhaven National Laboratory. These representatives presented programs in their respective institutions, and several common themes emerged. The programs frequently focus on inquiry-based training, hands-on experiences, or laboratory research to strengthen teachers' content knowledge and familiarity with performing research. DOE uses its national laboratories as a resource both for teachers and students in these efforts. A common theme from these discussions was that assessing the effectiveness of educational activities remains challenging, although programs can make progress by relying on standardized instruments and by teaming with evaluation experts.

The second half of Session 2 presented publicly funded outreach programs considered representative of exemplary programs. The presenters were **Irwin Talesnick**, Queens University; **Constance Blasie** and **Michael Klein**, University of Pennsylvania; **Sergey Nizkorodov**, University of California-Irvine; and **Gil Pacey** of Miami University, Ohio. Outreach methods included the ChemEd conferences, summer workshops, and masters' programs for teachers—all with varying levels of evaluation. Some programs and workshops offered course credit as a method to increase teacher involve-

ment. A common theme was to have teachers that complete these programs impart what they have learned to their peers, multiplying the number of people reached. A poster session, which contained a sampling of teacher outreach programs, followed these discussions.

One the second day, Session 3 focused on privately funded outreach programs. Speakers were **Bridget McCourt**, Bayer Corporation; **Reeny Davison**, ASSET program; **Bryce Hach**, Hach Scientific Foundation; **Patricia Soochan**, Howard Hughes Medical Institute; and **Sandra Laursen**, University of Colorado-Boulder. These programs have a broad range, with some focusing on elementary and middle school education and others on high school education. They seek to generate future research chemists, chemistry teachers, and a scientifically literate public through a variety of methods, including volunteerism, workshops, educational materials, and scholarships. Some programs perform very little evaluation, while some place a great emphasis on it. The workshop ended with a panel to consider what actions could be useful in the future. These suggested actions were the opinions of individual panel members and do not represent consensus recommendations. The panel was moderated by Bill Carroll and included **Joan Prival**, **Mary Kirchhoff**, **Penny J. Gilmer**, **Gerry Wheeler**, and **Hai-Lung Dai**. The panel discussed possible improvements in coordination, program evaluation, and a focus on the early stages of education as a part of a comprehensive effort to improve U.S. science education.

2

Science and Science Education in the United States

Major Points in Chapter 2

Research and development (R&D) have become more internationally distributed even as R&D in the United States has grown substantially in scale and scope.

Despite recent increases in some measures of scientific and mathematical proficiency, U.S. students on average still lag behind their international counterparts in some areas, and major gaps persist between groups in the U.S. population.

Opportunities to participate in practice teaching and professional development are unevenly distributed and insufficient to transform the knowledge and skills of teachers as a whole.

Parents recognize the need for improved science and mathematics education but tend to be satisfied with the amount of science and mathematics their own children study in school.

Students and parents in other countries tend to associate success in science and mathematics not with innate talent but with the effort invested in those subjects.

The National Science Board has identified better coordination and more effective teaching as the greatest needs of the U.S. educational system.

Every two years the National Science Foundation releases a new edition of its *Science and Engineering Indicators*. Kathryn Sullivan, director of the Battelle Center for Mathematics and Science Education Policy at Ohio State University, led off the workshop by presenting some of the data from the 2008 *Indicators* that are especially pertinent to science and science education in the United States.[1]

---

[1] National Science Board. 2008. *Science and Engineering Indicators 2008*. Arlington, VA: National Science Foundation.

THE R&D ENTERPRISE

Based on 2002 data, the amount spent on research and development (R&D) by European nations as a whole and by Asian nations as a whole nearly matched the amount spent in North America, which represents a significant expansion of R&D expenditures in Europe and Asia. "The R&D enterprise is becoming more of a shared enterprise," said Sullivan, with "greater competency, greater depth, and greater expenditures in more regions now than ever before in history." However, there have been some notable exceptions to the general increase in R&D expenditures outside the United States. Japanese expenditures, for example, which increased dramatically in earlier years, stagnated in the early 2000s.

In 1980, approximately 31 percent of the people in the world with education beyond high school lived in the United States. By 2000, that number was down to 27 percent. During that period, China's share of the total increased from 5.4 to 10.8 percent, while Japan's share dropped from 9.9 percent to 6.4 percent. The United States still has a greater absolute number of people with college education than any other country, but Asian countries as a whole are rapidly increase their percentage of the total.

Similarly, universities in the United States still award more doctoral degrees in the natural sciences and engineering than those in any other country—a total of more than 20,000 in 2005 (Figure 2.1). However, the number awarded in China has grown very rapidly since the early 1990s, while the number of Ph.D.s awarded in other countries has been stable or slowly rising.

Federal R&D expenditures have risen substantially in the United States since 1980, driven partly by a major increase in spending on health-related R&D (Figure 2.2). Defense R&D, which is predominantly development funding, also has risen dramatically over that period, with an especially sharp increase since the terrorist attacks of 2001.

The growth of employment in science and engineering fields has outpaced job growth in other sectors of the economy for decades (Figure 2.3). Furthermore, the earning power of workers with science and engineering degrees, regardless of the area in which they work, is higher than for their peers in other areas. The earning potential of jobs in

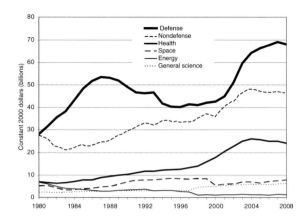

FIGURE 2.2 Federal R&D in billions of constant 2000 dollars has fluctuated since 1980, with a sharp increase in the first years of the new century. SOURCE: National Science Board. 2008. *Science and Engineering Indicators 2008*. Two volumes. Arlington, VA: National Science Foundation (volume 1, NSB 08-01; volume 2, NSB 08-01A).

science and technology runs counter to what she sometimes hears from students, Sullivan said. Students "seem to have a sense that these are not good-earning jobs—and perhaps they're not, compared to financial services or top-tier jobs on Wall Street. But against the broad backdrop of the U.S. labor pool, [there are] sustained job growth and better financial prospects for graduates with science and engineering degrees than for those who lack [them]. The data are quite clear and strongly sustained over many decades in that regard."

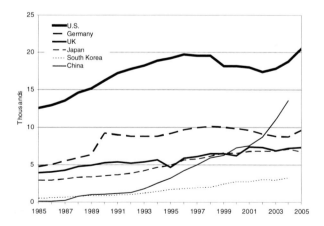

FIGURE 2.1 The United States and China now account for the largest number of doctoral degrees awarded in the natural sciences and engineering. SOURCE: National Science Board. 2008. *Science and Engineering Indicators 2008*. Two volumes. Arlington, VA: National Science Foundation (volume 1, NSB 08-01; volume 2, NSB 08-01A).

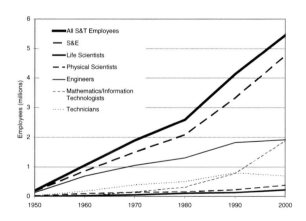

FIGURE 2.3 Science and technology employment has risen dramatically in the past half century. SOURCE: National Science Board. 2008. *Science and Engineering Indicators 2008*. Two volumes. Arlington, VA: National Science Foundation (volume 1, NSB 08-01; volume 2, NSB 08-01A).

People with science and engineering degrees work in every major sector of the economy: management, finance, information, government, education, manufacturing, and various technical services. Even among employees with doctoral degrees, only 44 percent work in college and university settings, while 33 work in the for-profit sector. Also, the number of jobs in all sectors of the economy that require the equivalent of a bachelor's degree in science or engineering is growing. People with science and engineering degrees are not confined to laboratories or professional service firms, Sullivan said. "There's good employment throughout the economy."

K-12 EDUCATION

Among the new data available in the 2008 *Indicators* are longitudinal data on the concepts and skills that students master as they move through the early grades. For example, more than 90 percent of fifth graders are proficient in multiplication and division, but only about 40 percent are proficient in rates and measurements, while the proficiency level for fractions in the fifth grade is barely above 10 percent. Among twelfth graders, 96 percent are proficient in simple arithmetic operation on whole numbers, but only 79 percent are proficient in simple operations with decimals, fractions, powers, and roots, and only 4 percent are proficient in solving complex multistep word problems.

The overall proficiency level in mathematics has been climbing in the fourth and eighth grades since 1990, but it has been stable at those grades in science, and twelfth-grade proficiency in science has fallen somewhat since 1996 (Figure 2.4). "We're not making the progress that we claim to and are working to make when it comes to science proficiency," said Sullivan.

Boys and girls start kindergarten at about the same level overall in mathematics performance. By the end of fifth grade, average boys' gains are greater than girls' by a small margin. With respect to race and ethnicity, average performance gaps already exist in kindergarten and widen across the full span of grades. By fifth grade, the average score for a black student is equivalent to the average score for a white third-grader. Students with mothers who have higher levels of education start kindergarten with higher scores than students whose mothers have less education, and these gaps also increase through fifth grade. The same observation is seen for families with incomes below the poverty line compared to families above the poverty line.

One interesting finding is that these gaps correlate strongly with loss of learning during the summer. Lower- and upper-income students make similar gains during the school year, but lower-income students experience sharper declines in performance over the summer while upper-income students do not fall back as sharply.

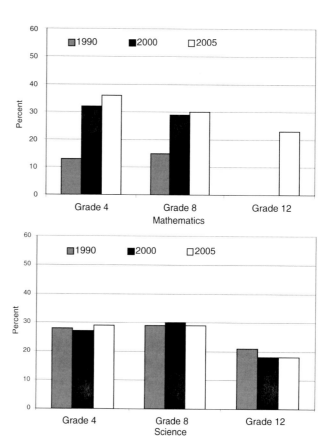

FIGURE 2.4 Proficiency in mathematics has risen for fourth and eighth graders since 1990 but has remained largely stable in science. SOURCE: National Science Board. 2008. *Science and Engineering Indicators 2008.* Two volumes. Arlington, VA: National Science Foundation (volume 1, NSB 08-01; volume 2, NSB 08-01A).

As in mathematics, new tests in science measure different aspects of proficiency, such as making inferences, understanding relationships, interpreting scientific data, forming hypotheses, developing plans, and investigating specific scientific questions. Again, boys show slightly higher average scores in third grade and maintain a small difference through fifth grade. By third grade, white and Asian-American students are higher in average score than African Americans and Hispanics, and by fifth grade, none of these gaps have narrowed.

The numbers of high school students who have taken specific courses in most of the sciences and engineering have grown since 1990 (Figure 2.5). For example, the number taking chemistry rose from 44 percent in 1990 to 55 percent in 2000 before falling off slightly in 2005. Similarly, the number taking advanced biology (where "advanced" is defined as courses that not all students are required to take) rose from 26 percent in 1990 to 39 percent in 2005. This is an important trend, said Sullivan, because taking advanced

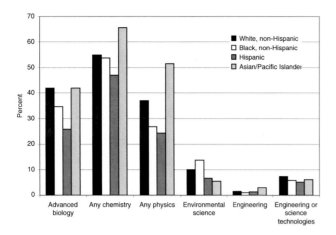

FIGURE 2.5 The percentages of high school graduates who have taken advanced science and engineering courses have increased since 1990. SOURCE: National Science Board. 2008. *Science and Engineering Indicators 2008.* Two volumes. Arlington, VA: National Science Foundation (volume 1, NSB 08-01; volume 2, NSB 08-01A).

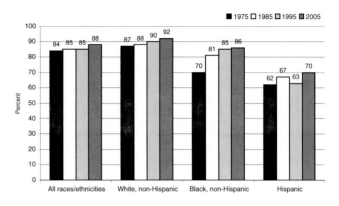

FIGURE 2.6 The high school completion rates of minorities have increased substantially since 1975. These rates measure the percentage of 18- to 24-year-olds who are not enrolled in high school and hold a high school diploma or equivalent credential such as a general equivalency diploma certificate. SOURCE: National Science Board. 2008. *Science and Engineering Indicators 2008.* Two volumes. Arlington, VA: National Science Foundation (volume 1, NSB 08-01; volume 2, NSB 08-01A).

courses is correlated with higher rates of college enrollment, higher rates of success in first-year college courses such as college algebra, and a greater likelihood of further workforce training. Completing advanced mathematics classes in high school also is directly associated with women's majoring in mathematics and science in college at higher rates. "An important leverage point if we want to move more women into chemistry or any other field in college is to be sure that we're working hard on giving them a strong core curriculum in math, certainly in high school, and setting them up strongly for that in middle school."

High school completion rates of 18- to 24-year-olds increased from 84 to 88 percent from 1975 to 2005 (Figure 2.6). These rates went up much more for African Americans—from 70 to 86 percent—and for Hispanics—from 62 to 70 percent—than for other groups over that period. Yet "this is one of those indicators that becomes striking when we place ourselves in the international comparison," Sullivan said. When compared with 22 other OECD (Organisation for Economic Co-operation and Development) countries using similar measures of graduation rates—with Norway first and Mexico last—the United States is sixth from the bottom.

Finally, in international comparisons of mathematical and scientific proficiency, U.S. students do quite well in the fourth grade. Eighth graders are still holding their own with respect to mathematics and science. However, high school students do markedly less well in international comparisons, especially when tests measure the ability to apply knowledge gained in school to less familiar problems.

TEACHERS

The 2008 *Indicators* has several new types of information about the preparation and quality of teachers, including information on pre-service education, practice teaching, degree attainment, and certification status.

Mathematics and science teachers with fewer than five years of teaching experience who report having practice teaching opportunities were more likely than teachers who did not have practice teaching opportunities to have learned about different pedagogical techniques, such as assessing students and using a wide variety of instructional materials. The percentage of teachers who report having done practice teaching is inversely related to the concentration of minority and poor students in schools, so the teachers of minority and poor students are less likely to have engaged in practice teaching. "We are shorting our students and giving our early career teachers a much harder hill to climb when we don't give them practice teaching opportunities," Sullivan said.

New indicators also show that more than 90 percent of teachers report participating over the past year in professional development activities consisting of short-duration workshops, conferences, and training seminars (Figure 2.7). Yet many years of evidence show that more than 40 to 50 hours of professional development and continuity of training are essential to have an effect on teacher practice, teacher competency, and improving a teacher's content knowledge, Sullivan observed. In contrast, only about a third were able to access university courses related to teaching, and about the same percentages were able to engage in research on a topic of interest. Also, teachers report participating in an average of just 32 hours of subject matter professional development.

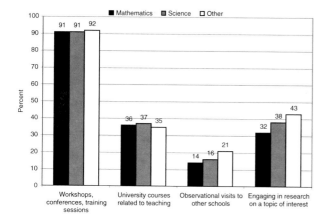

FIGURE 2.7 More than 90 percent of public middle and high school teachers report participating in workshops, conferences, and training sessions over the past 12 months, with smaller percentages participating in other professional development experiences. SOURCE: National Science Board. 2008. *Science and Engineering Indicators 2008.* Two volumes. Arlington, VA: National Science Foundation (volume 1, NSB 08-01; volume 2, NSB 08-01A).

"So we're only about halfway there in terms of the number of hours we probably should be providing to our teachers to improve skills in the classroom."

Sullivan made a personal plea to focus on the middle school years. Scientists in colleges and universities tend to focus on high school because courses organized by discipline appear at that level, but "middle school, in my opinion, is the sweet spot." Those are the years when children decide whether they are good at mathematics and science and start making choices accordingly. Science and mathematics teachers in high school tend to have more skills than the corresponding teachers in middle schools, depending on the socioeconomic status of the school. "I urge you to think about ways that you can . . . coordinate efforts to impact the preparation and skills of middle school teachers," she said. Local scientific societies, science museums, colleges and universities, and technology-based businesses all can help improve the skills, knowledge, and confidence of middle school teachers. "You'll have more kids to impact in high school if you can have a leverage effect at the middle school grades."

CURRICULUM STANDARDS

States recently have been improving their standards for K-12 education and bringing a better consistency to reviewing and revising these standards—despite periodic calls for teaching intelligent design creationism in science classrooms. Yet just slightly more than half of the states require three or more years of both mathematics and science for high school graduation, even though many national reports and organizations have identified that amount of mathematics and science as an essential core curriculum in the subjects.

In national surveys, 67 percent of Americans say that greatly increasing the number and quality of mathematics and science courses would improve high school education, and 62 percent say it is crucial for most students to learn higher mathematics skills. Yet when parents are asked about their own children, they tend to say that they are satisfied with the amount of mathematics and science they study in school. Furthermore, today 20 percent fewer respondents feel that children are not taught enough mathematics and science than in 1994.

One difficulty, said Sullivan, is that education in the United States is the equivalent of a radically divided market because it takes place in 16,000 largely autonomous school districts. If a new drug is developed by a pharmaceutical company, it can be approved by the Food and Drug Administration (FDA) and then sold to everyone in the United States. But "if you have figured out the equivalent of . . . the latest cure in science education, you must, in a sense, persuade 16,000 FDAs that this is the antidote to their ills. And therein lies one dimension of what makes this problem so massive."

PUBLIC ATTITUDES AND EXPECTATIONS

Polls show that the general public's primary source of information about scientific issues is the Internet, followed by television, books, and magazines and newspapers (Figure 2.8). The Internet and television have been capturing the attention of rising percentages of the public, while books have fallen precipitously (the proportion of those who say that books are their primary source of scientific information decreased from more than 20 percent in 2001 to less than 10 percent in 2006).

Knowledge of scientific facts and processes among the general public in the United States correlates closely with attitudes toward science. People who know more about science tend to believe that it has a positive role in society and has the potential to contribute to the public good.

With regard to student attitudes, confidence in being able to do mathematics or science correlates positively with achievement within countries. Yet across countries, confidence is negatively correlated with achievement. In other words, U.S. students with reported higher levels of confidence scored lower than students in other countries who reported themselves to be less confident. For example, 39 percent of U.S. students said that they usually do well in mathematics, while just 4 percent of students in Japan said the same. Yet the average mathematics score in the United States was considerably lower than that in Japan.

"We tend to believe that a large determinant of a student's success in school is ability," Sullivan said. "In many other

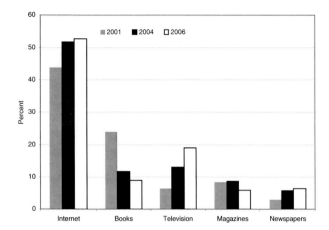

FIGURE 2.8 Use as the primary source of information about specific scientific issues for Americans has decreased for books and has increased for the Internet and television. SOURCE: National Science Board. 2008. *Science and Engineering Indicators 2008.* Two volumes. Arlington, VA: National Science Foundation (volume 1, NSB 08-01; volume 2, NSB 08-01A).

cultures and countries, it is perceived and believed that the correlation is with effort, with work, with investment, and self-discipline."

THE LOSS OF LEADERSHIP

In the 2007 report *Rising Above the Gathering Storm: Energizing and Employing America for a Brighter Economic Future,*[2] a committee of prominent national leaders stated, "We fear the abruptness with which a lead in science and technology can be lost—and the difficulty of recovering a lead once lost, if indeed it can be regained at all." Sullivan pointed out that there remains a very sharp divide between the perceptions of leaders in the United States and the perceptions of parents. Even though most parents know that more mathematics and science are needed for students, they tend not to extend that reasoning to their own children.

The National Science Board (NSB), which Sullivan vice chairs, recently reviewed the past three decades of reports on education in science, technology, engineering, and mathematics (STEM) and, responding to a congressional request, developed a plan to "convert all those grand words into some forward action." The board's report, *National Action Plan for Addressing the Critical Needs of the U.S. Science, Technol-*

ogy, Engineering, and Mathematics Education System[3], says that the nation faces two central challenges in constructing a strong and coordinated STEM education system. The United States needs to achieve greater coherence in STEM learning, including enhanced horizontal coordination and vertical alignment among educational systems. The nation also needs to ensure an adequate supply of well-prepared and highly effective teachers.

The NSB directed attention to issues that influence the quality of the teaching force in the United States, including compensation, stronger pre-service and in-service teacher education, increased teacher mobility between districts, and greater commonality of national teacher certification standards. The board recommended establishing an independent, non-federal, national council that could coordinate and facilitate STEM programs and initiatives throughout the nation while also informing policy makers and the public about the state of STEM education in the United States. The council would include representatives from foundations, higher education, business and industry, state and local governments, Congress, the executive branch, STEM educators, disciplinary scientists, informal STEM educators, and other organizations. The council could strengthen the linkage between high school education and the workforce, in part by working with K-16 STEM-focused councils in each state.

---

[2]National Research Council. 2007. *Rising Above the Gathering Storm: Energizing and Employing America for a Brighter Economic Future.* Washington, DC: The National Academies Press.

[3]National Science Board. 2007. *National Action Plan for Addressing the Critical Needs of the U.S. Science Technology, Engineering, and Mathematics Education System.* Arlington, VA: National Science Foundation.

3

The High School Chemistry Teacher: Status and Outlook

Major Points in Chapter 3

Longitudinal data demonstrate that exposure to particular subjects in high school chemistry, frequent peer interactions, and studying high-level mathematics are positively associated with chemistry grades in college, while time spent on community and student projects, labs, and instructional technologies can be negatively associated with college chemistry grades.

Most high school chemistry teachers have taken college courses above the level they are assigned to teach, but they report needing help using technology in science instruction, teaching classes with special needs students, and using inquiry-oriented teaching methods.

Laboratories in high school chemistry tend to be disconnected from coursework, to focus on procedures rather than on clear learning outcomes, and to provide few opportunities for discussion or reflection.

New requirements that high school students take more advanced science courses have increased the need for well-prepared chemistry teachers.

A major challenge for high school chemistry teachers is connecting the subject to everyday experiences, and professional development that focuses on this linkage can be especially valuable.

High school teachers can have a tremendous impact on students' interest and performance in the sciences. Many scientists talk about an especially inspiring teacher they had in high school. High school teachers often report that former students have told them about successes in college that they attribute to experiences in that teacher's class. "There's very little doubt in anyone's mind that teachers can, conceivably, have a tremendous impact on students' interest and performance in the sciences," said Robert Tai, an associate professor in the Curry School of Education at the University of Virginia.

Yet how can anyone know that this kind of anecdotal evidence is representative? Only broad-based representative sampling can provide solid data about the effects of high school science classes in general, Tai pointed out. Without such data, several important questions are left unanswered. How pervasive is teachers' influence? Are some teaching practices more effective than others? Can teachers' influence span the years from high school to college?

The data needed to answer these questions must be drawn from many students and classes, be representative of students, and in many cases, extend over periods of years.

Ideally, such data would include information about what students were doing when they were very young and what they were doing in college. The questions asked of students need to be specific enough to determine why they made the choices they did, and the people who are answering the questions need to care enough about the project to provide thoughtful responses.

Tai and his colleagues have used three different data sets to explore these issues. The first is the National Education Longitudinal Study (NELS), which has been collecting data from a sample of several thousand students since 1988. The second is Project FICSS—Factors Influencing College Science Success—a national survey of introductory college science students in biology, chemistry, and physics in which 67 colleges and universities have participated. The third is Project Crossover, which is a nationally representative survey of approximately 3,000 chemists and physicists and 1,000 graduate students in those disciplines.

One of the questions on the NELS questionnaire has been, What kind of work do you expect to be doing when you are 30 years old? In a study published in *Science* in 2006, Tai and his collaborators combined the answers to this question by eighth graders with data on factors such as demographic indicators, school attendance, and results on standardized achievement tests.[1] They asked whether an eighth grader who expressed an interest in a science-related career was more likely to graduate college with a degree in science. As expected, they found that students who said they wanted to be in a career related to the life sciences, physical sciences, or engineering were two to three times more likely to earn a degree in that area than students who did not express this interest.

They also found that eighth graders who performed higher on standardized tests in mathematics were more likely to graduate with a degree in the sciences or engineering. However, their analysis produced an unexpected result. Eighth graders who are interested in science or engineering but with average mathematics scores are more likely to graduate with a college degree in those fields than the eighth graders who scored highest in mathematics. "That means that there is some indication that it's not all about test scores, especially not in the eighth grade," said Tai. Yet those eighth grade scores are used to track students into mathematics classes when they enter high school, meaning that some students with an interest in science and engineering could be tracked into high school classes that make it difficult or impossible for them to achieve their goals. "We should take a very close look at how we go about doing this kind of thing. . . . That's an approach that we really need to reevaluate."

One of the questions in the Project Crossover survey asked practicing chemists, physicists, and graduate students

in those fields when they first became interested in science. About 70 percent of both groups reported developing an interest in science in grades K-8, about 24 percent reported developing their interest in grades 9-12, and about 6 percent reported developing their interest in college (Figure 3.1). However, when asked when they developed an interest in their "career discipline," the results were somewhat different. The survey showed that 29 percent of scientists and 23 percent of graduate students reported developing their interest in grades K-8, 52 percent of scientists and 56 percent of graduate students reported developing their interest in high school, and 18 percent of scientists and 21 percent of graduate students reported developing their interest in college (Figure 3.2). Thus, "you can't ignore any particular level," Tai said. To realize the full potential of the workforce to understand and contribute to science, the subject needs to be emphasized at each grade level.

Tai and his collaborators also have looked specifically at the factors that contribute to success in chemistry in college, as measured by the grades received in their college chemistry classes. They investigated instructional practices, key content and concepts, lab experiences, the use of technology, and student projects. They then constructed comprehensive models of the connections between these factors and college performance in both physics and chemistry. "What we're finding is that there is a connection, and it's robust and fairly consistent from sample to sample." Yet the connection also leads to some surprising and counterintuitive conclusions about high school science classes.

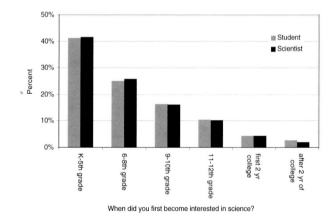

When did you first become interested in science?

FIGURE 3.1 The majority of graduate students and scientists report becoming interested in science in elementary and middle school, but about 30 percent develop their interest in high school and college. SOURCE: Tai, R. H. 2008. Research on Student Interest and Performance: Factors within a Teacher's Influence. Presentation to Chemical Sciences Roundtable, Washington, DC, August 4, 2008. Based on results of Tai, R. H. and F. Xitao. Project Crossover: A Study of the Transition from Science Graduate Student to Scientist, (NSF grant REC 0440002). University of Virginia.

[1]R. H. Tai, C. Q. Liu, A. V. Maltese, and X. Fan. 2006. Planning early for careers in science. *Science* 312(5777):1143-1144.

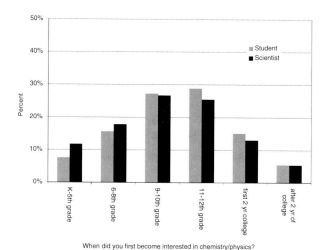

When did you first become interested in chemistry/physics?

FIGURE 3.2 Slightly more than half of graduate students and scientists report becoming interested in their career discipline during high school, but significant fractions do so both earlier and later than high school. SOURCE: Tai, R. H. 2008. Research on Student Interest and Performance: Factors within a Teacher's Influence. Presentation to Chemical Sciences Roundtable, Washington, DC, August 4, 2008. Based on results of Tai, R. H. and F. Xitao. Project Crossover: A Study of the Transition from Science Graduate Student to Scientist, (NSF grant REC 0440002). University of Virginia.

One conclusion is that the inclusion of particular concepts in high school chemistry has an impact on student performance in college.[2] For example, students who received recurring exposure to the subject of stoichiometry did better than students who had no exposure to the subject. Yet students who had recurring exposure to nuclear reactions did worse in college chemistry than did students who had no high school exposure to the subject at all. The reason seems to be that nuclear reactions are among the last topics covered in the high school chemistry curriculum, which means that teachers have to be speeding through the material to get to it, according to Tai. "You're flying along too fast, basically, covering way too much stuff. That's what this is indicating."

In a study published in 2007, Tai's research group looked at the connection between instructional practices and grades in college chemistry classes.[3] Surprisingly, they found that the more demonstrations students observe in high school chemistry, the worse they do in college. Tai speculated that too many demonstrations might be "dog-and-pony shows"

that focus on the demonstrations themselves and not on what the demonstrations mean.

Having students interact with each other in high school chemistry classes, as opposed to having them work individually, positively affects their performance in college. Yet time spent preparing for standardized tests has a negative effect. Time spent on community and student projects also has a negative effect on grades, especially for the weaker students in high school chemistry. Community and student projects may in general be too open-ended, Tai observed. Such projects can have relatively little structure, which may not be a problem for high-performing students, but "the ones who are struggling in school . . . typically are struggling to understand what's going on and need to have some kind of structure in their learning." Especially for students without a preexisting understanding of and interest in science, projects may have to be combined with robust content instruction. Furthermore, different students may need different kinds of instruction. "The same science doesn't fit all students," said Tai. "Different approaches work better for some students versus others."

Having large numbers of laboratories in high school chemistry is negatively associated with grades in college.[4] That does not necessarily mean that all labs are bad, Tai cautioned. "It may well be that loading students up on these hands-on experiences without the kind of debriefing that's necessary to help them understand what it is that they're doing in the labs isn't that helpful." Similarly, time spent preparing for and understanding lab procedures had a negative effect on college grades. These findings have not been very popular with chemistry educators, Tai admitted, yet they can reveal some important aspects of the interaction between students and the content of a high school chemistry course.

In work that was still unpublished at the time of the meeting, Tai examined the effect of instructional technology in high school chemistry courses on college performance in chemistry classes. Although many billions of dollars have been spent on instructional technologies in high schools, students who use these technologies frequently in their chemistry classes do *worse* overall in college. "This is a bit distressing, given the amount of money that we're spending," Tai said. Yet it may be an indication that "mainly what we're doing is asking teachers to fit their teaching to the technology and not so much fitting the technology to the teacher. It may well be that we're proliferating technology faster than the teachers are able to incorporate it into what it is they are doing.

District- or school-level policies also can affect high school chemistry instruction with a corresponding influence on college chemistry performance. For example, lengthening

[2]Tai, R. H., Ward, R. B., and Sadler, P. M. (2006). High school chemistry content background of introductory college chemistry students and its association with college chemistry grades. *Journal of Chemical Education* 83(11):1703-1711.

[3]Tai, R. H., and P. M. Sadler. 2007. High school chemistry instructional practices and their association with college chemistry grades. *Journal of Chemical Education* 84(6):1040-1046.

[4]R. H. Tai, P. M. Sadler, and J. F. Loehr. 2005. Factors influencing success in introductory college chemistry. *Journal of Research in Science Teaching* 42(9):987-1012.

of high school class periods from 45 or 50 minutes to an hour and a quarter or an hour and a half did not make much difference to college grades.[5] Class size does have an influence, but only if the classes are very small—10 students or fewer—and the only state that has classes that small is Vermont, where high school class size averages about 11.[6]

Finally, Tai's group has looked at whether taking a different science class in high school improves grades in college chemistry.[7] No such effect was observed for either biology or physics classes. However, taking calculus in high school had a big effect not only on chemistry grades but on college physics and biology grades as well. "I don't think it's necessarily the content," said Tai. "It may well be the type of reasoning and understanding that's required, the organization of thought that's required to progress in mathematics."

At the end of his talk and again during the question-and-answer session, Tai discussed the nature of the link between teaching practices in high school chemistry classes and grades in college chemistry. His results are all associational, he cautioned, making it difficult to draw causal links from any given practice or action to an outcome. "But the fact that the same students were followed for this period of time gives us more of a basis to draw conclusion about whether [a given practice] is important."

AN OVERVIEW OF HIGH SCHOOL CHEMISTRY TEACHERS

There are between 30,000 and 40,000 high school chemistry teachers in the United States, according to informed estimates discussed at the meeting. Pinning down an exact number is difficult because many teachers are engaged in teaching that is out of their fields. According to data gathered in 2000, slightly more than half of all high school chemistry teachers in the United States are female, 91 percent are white, roughly half have a master's degree, and approximately one-third may be reaching retirement in the next 10 years.[8] The data may be somewhat old, said Gerry Wheeler, executive director of the National Science Teachers Association, but they probably still capture fairly well the demographics of high school chemistry teachers. The most surprising statistic to him, said Wheeler, is that roughly one-third of those who teach chemistry have three or more preparations a day,

---

[5]K. M. Dexter, R. H. Tai, and P. M. Sadler. 2006. Traditional and block scheduling for college science preparation: A comparison of college science success of students who report different high school scheduling plans. *High School Journal* 89(4):22-33.

[6]V. L. Wyss, R. H. Tai, and P. M. Sadler. 2007. High school class size and college performance in science. *High School Journal* 90(3):45-53.

[7]P. M. Sadler and R. H. Tai. 2007. The two high-school pillars supporting college science. *Science* 317(5837):457-458.

[8]Smith, P. S. (2002). The *2000 national survey of science and mathematics education: Status of high school chemistry teaching.* Chapel Hill, NC: Horizon Research, Inc.

TABLE 3.1 College Science and Education Courses Taken by High School Chemistry Teachers

Course	Percentage
General methods of teaching	90%
Methods of teaching science	73%
Supervised student teaching in science	66%
Instructional uses: computers/other technologies	42%
General introductory chemistry	99%
Organic chemistry	93%
Analytical chemistry	68%
Physical chemistry	51%
Biochemistry	47%
Other chemistry	44%
Quantum chemistry	16%

SOURCE: Smith, P. S. (2002). The *2000 national survey of science and mathematics education: Status of high school chemistry teaching.* Chapel Hill, NC: Horizon Research, Inc.

including different levels of chemistry, other science courses, or courses outside science.

According to the same data source, 99 percent of high school chemistry teachers have completed a college course in general introductory chemistry, 93 percent have done so for organic chemistry, 68 percent have had analytical chemistry, and 51 percent have had physical chemistry (Table 3.1) "Most have had courses above the level they've been assigned to teach," said Wheeler. "That's not true in middle-level and elementary math and science. There are many times in elementary math and science where teachers are actually, shocking as it is, teaching material that is higher than the level they took in their preparation."

More than half of chemistry teachers say that they need help in using technology in science instruction, teaching classes with special needs students, and using inquiry-oriented teaching methods. Between one-third and one-half of teachers report needing help in understanding student thinking in science, learning how to assess student learning in science, and deepening their own science content knowledge.

Chemistry teachers, like high school science teachers in general, also report low levels of participation in professional development that is specific to science teaching. Schools face a dilemma in that regard, said Wheeler. The needs of their teachers are so varied that they find it easier to hire a generalist to talk about motivation, management, or some other general subject rather than addressing the content needs of individual science teachers. Even science teachers need very different kinds of professional development. "The variation is so wide that I haven't found schools being able to solve that problem."

According to data gathered by the Council of Chief State School Officers, the proportions of students taking chemistry in high school range from 87 percent in Texas to 13 percent

TABLE 3.2 Percentage of Students Who Have Taken
Chemistry Courses in High School

State	2006 [a]	Change[b]
Texas	87	+32
Iowa	86	+18
South Carolina	82	—
District of Columbia	76	−1
Wisconsin	74	−2
Pennsylvania	69	+3
Louisiana	66	+12
Tennessee	66	+14
New York	62	−2
Ohio	62	+4
Arkansas	59	−6
South Dakota	58	+6
Indiana	56	−1
North Dakota	55	−5
North Carolina	54	−9
Wyoming	51	—
California	49	+10
Minnesota	49	−2
Mississippi	48	−10
Utah	48	+1
Missouri	48	−3
Michigan	43	—
New Mexico	41	0
Oklahoma	39	−1
Idaho	29	−12
West Virginia	13	−47

NOTE: The numbers range from 87 percent to 13 percent, with substantial changes both upward and downward between 1996 and 2006.

[a]Percentage.

[b]Percent change between 1996 and 2006.

SOURCE: Council of Chief State School Officers. (2007). *State Indicators of Science and Mathematics Education: 2007*. Washington, DC: Author. http://www.ccsso.org/publications/details.cfm?PublicationID=371.

in West Virginia (Table 3.2). Even within states, there is great variation in how many students take chemistry and other science classes in high school. "We've got 16,000 school districts making 16,000 different kinds of decisions."

A particularly problematic aspect of high school chemistry is its use of labs, Wheeler pointed out. "The general laboratory situation is pretty deplorable." He pointed to a recent National Research Council study that examined the current state and effectiveness of high school labs, their interactions with technologies and school policies, and possible alternatives to labs.[9] Though the report addressed all high school labs, its points are just as relevant when made about chemistry labs.

The report pointed out that there is no consensus on what the goals of labs are or should be. For more than 150 years, scientists and educators have assumed that labs are essential

[9]National Research Council. 2005. *America's Lab Report: Investigations in High School Science*. Susan R. Singer, Margaret L. Hilton, and Heidi A. Schweingruber, eds. Washington, DC: The National Academies Press.

to teaching science. Yet lab experiences have often been isolated from the general flow of science teaching. For example, Wheeler pointed out that in two of the universities he has been associated with, the chemistry labs were independent of the chemistry course in the sense that the two could be taken at completely different times, and the same disconnect between labs and course work often happens in high school. Instead of being integrated into what the teacher is trying to accomplish, the lab is isolated and independent.

Current labs have other negative characteristics. They tend to be focused on procedures rather than clear learning outcomes. They provide few opportunities for reflection or discussion. They do not integrate the learning of content with the processes of science; they do not reflect instructional design based on recent cognitive research; and future teachers are not exposed to good labs as models of experiential learning.

Labs provide a prime opportunity to teach students who will not become scientists about the nature of science, yet this remains one of the greatest failings of high school science classes. Wheeler said that he spends considerable time defending the high school biology curriculum from demands that creationist ideas be included in the science curriculum. If Americans had a better grasp of the nature of science, they would be less likely to call for the inclusion of religious ideas in science classrooms.

Major changes will be required in several areas to improve high school labs. Schools, districts, and states will need to support meaningful reform in the design and use of labs. Undergraduate science education will have to change, and state standards will need to change so as not to discourage teachers from dedicating the time needed for effective labs. For example, the skills being tested at the state level through "No Child Left Behind" have nothing to do with lab experience that students should have. The high-stakes tests that have been adopted by many states end up "valuing what we measure instead of measuring what we value," Wheeler said.

Current assessments are not designed to measure accurately the outcomes of lab experiences. Developing and improving these assessments is not easy and will be expensive. "It's a very challenging problem to assess student achievement in the things we actually value," said Wheeler. Even today, most assessments are unaligned with even the best standards, and most sets of state standards are far from optimal. Although the standards developed by the National Research Council and the American Association for the Advancement of Science are high quality, said Wheeler, the states have altered and in many cases expanded them. "We now have 50 different state standards that are filled with factoids."

Beyond the problem of labs, said Wheeler, the greatest challenge is at the middle school level. Teachers need to understand the subjects they are assigned to teach. Yet especially in middle school, teachers are often misassigned. "When I was in my first year of teaching, we joked and said, 'Don't hum when you're walking down the hallway. The principal will turn you into the music teacher.'" Despite regulations at the state and district levels about teacher qualifications, inside an individual school, said Wheeler, "all bets are off."

Efforts to change education also must address the problem of scale. Teaching is a huge profession. There are 1.7 million teachers of science in the United States, including the 1.5 million elementary school teachers who are students' first teachers of science. Programs that reach just a few teachers may be important but cannot overcome the problems of scale that must be addressed. For example, sending underprepared middle school teachers back to college and university classes cannot raise their content knowledge enough for them to teach science well, especially given estimates that half of all chemistry teachers leave the profession within five years. Instead, the National Science Teachers Association (NSTA) has been working to create a large Web site that offers highly interactive four-hour engagements with science content designed for the adult novice learner. Another initiative has been to connect early-career teachers with experienced teachers through an electronic network. That effort has in turn led to the formation of the NSTA New Science Teacher Academy, which uses mentoring and other professional development resources to support science teachers during their initial years.

CHEMISTRY AT THE STATE LEVEL

Although no national organization representing chemistry teachers exists, many states have an active group for high school chemistry teachers. For example, the Associated Chemistry Teachers of Texas (ACT2), was created 27 years ago and is an affiliate organization of the Science Teachers Association of Texas. "It's not anywhere near the numbers we would like," said Roxie Allen, a former ACT2 president and a teacher at St. John's School in Houston, but "it's a very viable chemistry teacher group."

The issues associated with high school chemistry in Texas are representative of those that occur throughout the nation. The Texas Education Agency sets the content to be taught in science classes through the Texas Essential Knowledge and Standards, which include a large number of content requirements along with laboratory skills. Texas is also moving to a requirement that high school students take four years of science, which has caused the number of students taking chemistry in Texas to rise from 32 to 87 percent, with an ultimate goal of 100 percent. In addition, end-of-course exams are being instituted for chemistry and other subjects in Texas, with the chemistry exams now being field-tested. These changes in requirements have greatly increased the need for chemistry teachers in Texas. Previously, many high

school science teachers taught an integrated physics and chemistry course. These teachers now are being expected to teach physics and/or chemistry, creating a great need for additional teacher training. Many school districts also like to hire teachers trained in composite science, because they can teach many different science courses. Unfortunately, as a result of this broad training, these teachers often lack depth in individual subjects. Furthermore, many teachers are not adequately trained to supervise students in labs, and most teacher training programs do not focus on lab skills.

Many high school science teachers come through alternate certification routes from industry. They know the content well but have few teaching skills or little knowledge of how to teach. Mentoring can be a great help to them, but the quality of mentors differs greatly. Funding is also a very significant issue in Texas, with large disparities in the amounts different schools have to spend. For example, "most chemistry teachers do not have supply budgets," Allen observed. Class sizes can be very large: When Allen taught in public school, she had 37 students in one of her classes. "You cannot do a lab when you have 37 students and you have desks and lab tables for 24." More and more districts in Texas are adopting block scheduling, which is cheaper for schools. This results in less contact time with students and makes it harder for students to absorb the large amounts of material presented in a single class. Block scheduling is especially difficult for advanced placement (AP) courses, which are designed for daily classes.

The Science Teachers Association of Texas holds an annual Conference for the Advancement of Science Teachers. The two- to three-day program features many hands-on workshops, which "is critical," Allen said. "If you don't try something, then you're probably not going to do it when you get back to your classroom."

The ACT2 has a membership of about 800, with anywhere from 50 to 150 teachers coming to conferences held every other year. The group also has a very active e-mail network, part of which is devoted to employment assistance. In addition, it has local groups—the local group in the Houston area is the Metropolitan Houston Chemistry Teachers Association, which meets three to four times a year. This organizational structure enables extensive networking, Allen said. "I don't know that a lot of states have the kinds of opportunities that we have to communicate with each other."

ACT2 meets during the state conferences and during regional miniconferences of the Science Teachers Association of Texas. The president-elect of ACT2 hosts the ACT2 conference in a city near where he or she lives, generally at a local university or college. Costs are kept very low, so that more teachers can attend. The emphasis is on hands-on activities that teachers can take back to their classrooms. Teachers also have an opportunity to network with college and university professors from the area. Outside funding is important to some of these activities, so when funding changes or becomes less available, less can be done. Members of the group also interact with national groups, including the NSTA, and Texas hosted the ChemEd conference last year.

ACT2 has had trouble attracting members among younger teachers. The resulting loss of membership reduces funding, which dampens the number and scale of activities that can be conducted. "We have very few young teachers who are joining us," said Allen. With a third of teachers expected to retire in the next 10 years, "organizations like ours are going to be impacted seriously."

In response to a question about how to improve the funding of high school chemistry, Allen noted that teachers do not know whom to contact in industry to get financial support, even when a company may be willing to provide assistance. "It would be wonderful if industry and even academia would figure out a way to help high school teachers know how to get money to do things like workshops." Most teachers coming to workshops held by ACT2 are probably paying their own way, even though funding may be available to subsidize their attendance.

ON THE FRONT LINES

High school chemistry teachers generally enter the profession in one of three ways, said Caryn Galatis, who has been teaching in the Fairfax County public school system in the Virginia suburbs of Washington, DC for more than 30 years. A very small number come directly into teaching after finishing their bachelor's degrees. "In the last 20 years that I've been involved in hiring practices at my school, I think I've only interviewed three or four teachers that are directly out of undergraduate education."

The second route is that people work in industry for several years but find that they are unhappy and decide to try teaching. The third route is to switch careers from another profession into teaching.

In the State of Virginia, high school chemistry teachers cannot get a high school teaching certificate without an undergraduate degree in chemistry or the equivalent number of courses, though this is not a requirement for certification in all states. Virginia also requires five different education courses for certification, one of which is in science methods.

However, most of the people Ms. Galatis has interviewed were not certified. Instead, new chemistry teachers are hired with a provisional contract and are given three years to fulfill the education requirements for certification.

Also, retention of science teachers is difficult, she said, "partly because retention in teaching in general is difficult. People tend to stay three to six years and are out." High school chemistry teaching is especially difficult because teachers need to plan, teach, and manage their classes and also prepare for and run a laboratory program. "You're almost

preparing twice as much content and not given any more time to do it," said Galatis. Most teachers set up labs before and after school, and "there's a lot of time involved."

A lack of mentors also has a negative effect on retention. Even in the Fairfax County system, which provides more support for teachers than most systems, "very little support is given to new teachers who come into the profession, and I don't care whether they're young teachers or old teachers."

The disparities that exist between states also exist within states. Although Galatis teaches in one of the richest counties in the state, she also owns property in one of the state's poorest counties. In Fairfax County, more than 90 percent of high school students graduate with a credit in chemistry. In the county where she owns property, she estimates that the percentage is probably less than 40 percent.

The two factors that have had the greatest impact on science teaching during her career have been the "Science for All" movement and state exit exams, which Virginia instituted in the 1990s. When she began teaching, probably 30 to 40 percent of students took chemistry—mostly college-bound students who were interested in science. Now most students who intend to enter four-year colleges are expected to have taken chemistry. The movement toward Science for All has been implemented very differently in the State of Virginia. In Fairfax County, most students take four science credits in three different science areas, so most college-bound students take biology, chemistry, and physics. In other places in the state, students need three science credits to graduate, which they can do without ever taking chemistry or physics.

Nevertheless, many more students take chemistry now than in the past, which means that many chemistry students have very weak mathematics backgrounds. "You're teaching chemistry to students who don't necessarily have an interest in science. They're taking it because they need it to graduate, which changes greatly what teacher[s] need in their skill set in order to teach the complexity of chemistry." The greater diversity of students is especially a problem for older teachers who are within 10 years of retirement and do not necessarily have the skill sets to teach less prepared students.

Galatis said that she is a firm believer "that all kids can learn chemistry," but "they can't all learn it the same way." Younger teachers coming right out of their undergraduate education are much better prepared than are many older teachers to teach chemistry to a broad range of students, so "at least in the State of Virginia, I know the universities are doing a pretty good job with that population."

Galatis spends many hours after school mentoring other teachers, if she can get them to work after hours. Yet teachers wish they had more time to improve their skills. While many training sessions and other opportunities are available, they can be expensive and far away. Without this training, teachers are less able to show their students the excitement of chemistry through labs and other hands-on experiences.

Her school has recently made an effort to give its chemistry labs a much more practical base, so they do relatively few textbook labs. Instead, they do more content-specific labs that connect students with particular problems. "Doing a lab that makes kids see the connection between content that's hard for them, giving them that mental picture in their head, so it's not just memorization and textbook learning, is what's going to get kids to stay in science."

Brian Kennedy, who teaches at Thomas Jefferson High School for Science and Technology in the Virginia suburbs of Washington, DC, said that he got interested in chemistry in college, when a particularly inspiring organic chemistry teacher made him decide to major in chemistry. While in graduate school in chemistry, he began to meet people who had been involved in the Teach for America program. After a postdoctoral fellowship at the Army Research Laboratory in Maryland, he entered Teach for America. "I was probably quite an anomaly to go into Teach for America after 12 years of college-type work," he said.

After teaching in Houston, he began teaching in a rural area of North Carolina in one of the lowest-performing schools in the state. "It was an extremely challenging environment" marked by many long days and nights of teaching, coaching, and helping the students in his classes. "I was able to see firsthand the extreme difficulties that a lot of kids had beyond the classroom." Many of his students could not read at a high school level, much less take chemistry, "yet here they were in a chemistry class. It was an extreme challenge to get them where you want them to be to do well in chemistry."

Resources were virtually nonexistent—sometimes he had a computer and a printer but very few materials or supplies, and the computer had no access to the Internet. "It took me a long time and a lot of grant writing to get the materials I needed for how I wanted to teach."

After interviewing for a new teaching position, he ended up at the Thomas Jefferson High School for Science and Technology, which is one of the top high schools in the country. For the past six years, he has taught all levels of chemistry there, including organic chemistry with instrumental methods of analysis.

Even though he now teaches in a very different environment than before, "there still seems to be an issue of getting the financial resources you need to do things for the caliber of student you think you have. That's been a common thread anywhere I've taught."

Teachers need greater access to outreach programs, Kennedy said. Many colleges and government agencies have programs designed to help, yet there is a disconnect between the teachers and the programs. "If teachers themselves could be more involved with creating the outreach opportunities, they're the ones who are in the trenches and understand what the real issues are."

Funding for education is becoming increasingly tight given the status of the economy, and especially the housing market. In Fairfax County, the funds available for the school system depend heavily on the state of the housing market. As real estate prices drop, so does the funding for education. Corporations and government need to increase their support of education to make up for the shortfall, he said.

Finally, Kiara Hargrove from Baltimore Polytechnic Institute said that she was inspired by her high school chemistry teacher, but she wanted to pursue a career in the biomedical sciences. As a researcher, however, she found that she got much more enjoyment out of presenting papers and talking with people at meetings than working in the lab, so she decided to go into teaching, where she could interact with students and watch them move into their own careers.

She began teaching at the middle school level, which "is a very different beast than teaching just chemistry at the high school level." She was teaching all of the physical sciences, algebra, and later, biotechnology at a mathematics and science magnet school in Baltimore County. That experience allowed her to learn about and experiment with the methodology of teaching, she said, which was easier for her because she already knew most of the content. After six years she began teaching chemistry at the high school she had attended. Baltimore Polytechnic Institute is a mathematics, science, and engineering magnet school that is among the top schools in the State of Maryland. The students can take organic chemistry and biochemistry as well as AP chemistry.

Hargrove teaches health as well as chemistry. It is challenging, she says, to prepare for another course in a different discipline, but her experience in the biomedical sciences has made it easier for her to be enthusiastic about that assignment. The school has three positions for chemistry teachers: One teaches just chemistry; one teaches chemistry, health, and one other course; and the third teaches chemistry, organic chemistry, physics, and possibly environmental science. "The retention of that teacher is very hard," she said. "We've had a new teacher in that position for the past four years."

Chemistry is not one of the subjects that undergoes a major assessment in the State of Maryland. As a consequence, chemistry teaching is not a focus of the school's professional development activities. Yet the chemistry teachers feel that they need professional development opportunities, whether from the school, the district, or elsewhere.

The size of her classes varies from 30 to 39 students. Conducting labs is very challenging, she says, but "I try to figure out ways that I can get 39 students in a lab," even without an assistant. Sometimes she brings in her own materials, and sometimes she tries to do labs with everyday materials such as polyvinyl chloride (PVC) pipe. She says that she tries to make the labs correlate with the curriculum guide, even though the labs take longer than they do for other teachers when she uses them to engage in "meaningful conversations."

During the question-and-answer period, the three teachers emphasized the importance of using professional development opportunities to connect chemistry with the context of daily life. "That's where the kids really see the excitement and learning with chemistry and the sciences is when you put those two together," said Hargrove. These connections can help fulfill the mission statement of the chemistry teacher, which the panelists described as comprising the chemistry education of both future citizens and future scientists. According to Galatis, forging links across disciplines is also essential, both in teaching and among teachers. It can be hard to coordinate across curricula within a school, but this kind of coordination can be greatly beneficial for students and teachers alike.

When asked about their best professional development activity, Kennedy said that learning the basics of teaching were most important for him, since he already knew the content. For example, What will you do on the first day of class? If all 50 students have a piece of paper in their hands, what is the best way to collect those papers? "For new aspiring teachers, if you want to keep them in the classroom, professional development that would help them get through that first year would be a crucial step."

For Hargrove, the most valuable professional development has been how to differentiate instruction. Her students have extremely varied sets of skills. "You have to figure out how to address those students and address their needs." Also, many teachers have good content knowledge in chemistry but lack the communication and social skills to work effectively with students. "Professional development that addresses how to reach those students who may seem unreachable" is important.

Galatis said that the best professional development she has done has been run by universities or companies, especially when they provide an opportunity to learn a new technique or use new equipment. "Imagine trying to be in front of 20 to 30 kids doing a lab when you have never touched the equipment yourself. It's an impossible task to ask of teachers, and we ask teachers to do that in large numbers of ways." Professional development workshops also have their place because she can come away from them with ideas that can be readily applied in the classroom.

Galatis also said that teachers need help connecting the curriculum they are given with the practical day-to-day tools that are needed for students to understand concepts. "One of the biggest problems with chemistry teachers who actually have the content is that they never struggled with learning chemistry. They don't understand what these kids don't know." They need tools to help kids understand the concepts that are being presented.

4

Initiatives by Federal Agencies

Major Points in Chapter 4

The Science Education Partnership Award program at the National Institutes of Health relies on partnership organized around inquiry-based curricula to increase the scientific and health literacy of the U.S. public and to promote careers in the health sciences.

A variety of programs at the National Science Foundation, both in the Directorate for Education and Human Resources and elsewhere in the foundation, support the education and professional development of chemistry teachers.

Taking advantage of the laboratories it supports across the United States, the Department of Energy specializes in helping chemistry teachers and students gain hands-on experiences.

Though assessing the effectiveness of educational activities remains challenging, programs can make progress by relying on standardized instruments and by teaming with evaluation experts.

Many federal agencies have programs directed toward K-12 education that affect chemistry education either directly or indirectly. Several presenters at the workshop described examples of these programs and summarized available evidence about their impacts on teacher and students While the programs represent just a small sample of all the federal programs focused on K-12 education, they demonstrate many of the strengths and weaknesses of federal initiatives.

THE NATIONAL INSTITUTES OF HEALTH: PARTNERSHIPS FOR SCIENCE EDUCATION

Since 1991 the National Institutes of Health (NIH) has sponsored the Science Education Partnership Award (SEPA) program to increase the scientific and health literacy of the U.S. public and to promote careers in the health sciences among groups that have historically been underrepresented in those fields, including urban, rural, African-American, Hispanic, Native American, and female students. The program seeks to create partnerships organized around inquiry-based curricula between scientists or clinicians and educators, community organizations, and science centers. The underlying concept, said L. Tony Beck, a program officer at NIH's National Center for Research Resources (NCRR), is to avoid having "a clinician or a scientist in a university create something, take it to the school, and say, 'Trust us, this will work.'"

The topic of a SEPA project can be anything related to NIH-funded research. Past awards have focused on subjects ranging from aging to epidemiology to vaccine development to environmental health. Also, SEPA projects generally have to target state and local standards that are in place, so they not only augment the curriculum but in some cases can replace it.

Projects are funded at $250,000 for each of five years. Phase 1 consists of development and evaluation of the project, with a second phase involving the development, implementation, and evaluation of effective strategies for dissemination. The program increased in size substantially in the late-1990s when Representative John Porter essentially tripled the budget so that it could begin to serve science centers and museums. A wide variety of institutions have projects, including medical centers, universities, colleges, community colleges, research institutions, nonprofit organizations, and public and private school systems. NIH and NCRR directors have been very supportive of the program and have kept its budget stable, despite current pressures on NIH funding.

Different projects have different goals. Some are focused on students who would not normally think of going to college. Others, such as the projects run through science centers and museums, are designed to reach out to the broader public. For example, a SEPA project at Yale includes a family night that features an introduction to research on Lyme disease, the use of microscopes to look at various types of ticks, and an opportunity to enroll in clinical trials studying the disease.

The program currently has 60 to 70 projects, with a near-term goal of funding 80 projects. Many of the projects are in states that do not receive large amounts of NIH funding, which promotes the program's goal of increasing the public's understanding of and support for NIH-funded research. "We're gradually expanding the range of the SEPA program, [which] brings this kind of opportunity to a range of students and communities," said Beck.

Many of the SEPA projects have developed Web sites, some of which have won national awards. The national Web site (*http://www.ncrrsepa.org*) offers a way to search for programs by state, principal investigator, or educational focus. For example, searching on Washington, DC, leads to an exhibit on infectious diseases at the Marian Koshland Science Museum of the National Academy of Sciences.

Although the classroom projects supported by SEPA generally are focused on the biomedical sciences, students are exposed to a broad range of scientific and mathematical content as part of the lessons. Also, the learning is inquiry-based, so students engage in learning processes that are not necessarily present in typical science classes. They "have to learn to work in teams of kids they wouldn't normally hang out with," said Beck. "They have to write coherent sentences. They have to understand chemistry, statistics, math."

Teacher professional development is an important part of the SEPA program. More than half of the projects feature at least one or two weeks of professional development, typically during the summer. Many of the projects are in inner-city schools, and because the projects are for K-12 students, the teachers reached by the program include elementary and middle school as well as high school teachers. "Most of them didn't have much science in high school," said Beck. "They have to learn the content and they have to learn the pedagogical skills."

Evaluation of projects is also a critical part of the program. At least 10 percent of each budget needs to be spent on evaluation—$25,000 per year at minimum. Evaluation plans also have to be part of the original proposal, and the evaluation needs to be conducted during the entire duration of the project. External evaluators are required unless the use of an institutionally based evaluator can be justified. Furthermore, as the emphasis in the evaluation of educational programs has shifted toward greater reliance on quasi-experimental and experimental designs, the SEPA projects also have moved in this direction. Five years ago, very few projects had approval from institutional review boards (IRBs) for evaluation plans, which generally is required for randomized controlled trials or comparison group studies. Now approximately half of the classroom-based programs do have IRB approval. In addition, the focus of the last two annual meetings for the SEPA program has been evaluation.

Beck described a particular evaluation of a project by Nancy Moreno and her colleagues at the Baylor College of Medicine's Center for Educational Outreach. The evaluation tracked the performance of students in classes taught by teachers who attended one or more professional development sessions in the summers. The evaluation showed that it takes several years after the initial professional development experience before a substantial impact appears in the schools, and the impact is greater after several years of professional development sessions.

Considerable anecdotal evidence also points to the success of projects funded by the SEPA program, Beck said. As one student wrote following a session in a mobile laboratory that travels from community to community in a bus, "This is more fun than shopping at the mall because at the mall you can't have your DNA in a little tube."

THE NATIONAL SCIENCE FOUNDATION (NSF): IMPROVING THE EFFECTIVENESS OF TEACHERS

"Education—and particularly teacher education—is becoming a huge thing at NSF," said Joan Prival, a program director of NSF's Directorate for Education and Human Resources (DEHR). NSF programs for science teachers have five broad goals: increasing content knowledge, improving pedagogical skills, enabling teachers to engage all students, enhancing their conceptual understanding, and helping them retain understanding. NSF pursues these goals through a research and development effort designed to advance knowledge and to further the preparation of K-12 teachers of science, technology, engineering, and mathematics (STEM). In addition, NSF's programs encompass the entire continuum of a teaching career, from recruitment of new teachers to pre-service education to induction to continuing professional development. "It's a lifetime experience, and we have a number of programs that head into teacher education at various points in the continuum," said Prival.

Many previous NSF reports and programs have addressed the importance of meeting the demand for highly qualified STEM teachers, reducing attrition, and broadening participation in STEM teaching to groups that have been underrepresented in those fields. The programs are based in existing research and practice, including the knowledge and experience in teacher education built up at NSF over more than half a century.

Teacher education in STEM seeks to prepare K-12 teachers who are proficient in STEM concepts and topics, confident in their own grasp of STEM content, lifelong learners of this content, aware of rapidly changing disciplinary content, able to guide and assess STEM learning in age-appropriate ways, confident in the use of cyber-enabled tools, prepared to engage an increasingly diverse student population, and supported by STEM faculty, in collaboration with teacher education faculty and practitioners. This last point is especially important, said Prival, because STEM faculty at colleges and universities are a critical part of the education community; in fact, many DEHR programs require that STEM faculty be involved in educational projects.

Teacher education in STEM requires a research base that can serve as a foundation for improved teacher education models. For this reason, DEHR has a strong research program that addresses such issues as teacher preparation, induction, and professional development. The directorate also requires that teacher education programs have a rigorous evaluation component that measures outcomes in terms of increased production of well-qualified teachers, knowledge and dissemination of proven strategies, and evidence of a relationship between teacher education components and improved K-12 student learning. The directorate does not require that a certain percentage of a grant be spent on education. Rather, the evaluation component is part of the intellectual merit that is considered in reviewing a proposal. Evaluation also is connected to the dissemination plan included in a proposal, "because if you're going to be disseminating anything that's worthwhile, people are going to ask how do you know and what's the evidence."

Many DEHR programs address STEM teacher education, some directly and some more peripherally. The largest such program is the Math and Science Partnership program, which fosters collaborations between STEM faculty in colleges and universities and K-12 school districts. Teacher quality, quantity, and diversity are major objectives of the program. Individual projects, which incorporate evidence-based design and outcomes, feature challenging courses and curricula for teachers. Also, "these are large grants, and we expect institutional change at both the school district level and the institution of higher education," Prival said. In some ways, these projects draw inspiration from the teacher institutions NSF sponsored in the 1960s, but they want to do more than influence individual teachers. For one, the program is designed for teachers to become leaders in their districts. "They go to an intense residential, or in most cases semiresidential, institute for a couple of years, and then they go back to their school district as teacher leaders," said Prival. "They have all kinds of responsibilities in working with the other teachers in their building. And we're looking to impact student learning in the whole building, not just in the classroom of the teacher who participated."

A program designed to bring people into teaching who have very strong content knowledge is the Robert Noyce Teacher Scholarship program. The program supports people who are majoring in chemistry or other scientific disciplines in college to become K-12 teachers through scholarships and stipends. Future teachers agree that for every year of a scholarship they will work two years in a high-need district. The program also targets people who are changing careers by supporting their preparation in a teacher credentialing program. "It's not enough that they know the content," said Prival. "They need to learn how to convey that excitement and the knowledge that's associated with their field."

Discovery Research K-12 is another large program at NSF that supports research about, and the development and implementation of, innovative resources, models, and technologies for use by students, teachers, and policy makers. Assessment, public literacy, workplace readiness, and cyber-enabled learning are emphasized in the program.

A program that features resources specific to chemistry education is the National Science Digital Library (*http://nsdl. org*). In addition, the ChemEd Digital Library (*http://www. chem1.com/chemed*) has links to many chemistry education resources.

Other DEHR programs that have an influence on STEM teacher education include the NSF Graduate Teaching Fellows in K-12 Education (GK-12) program; the Advanced Technological Education (ATE) program; the Course, Curriculum, and Laboratory Improvement (CCLI) program; the Research and Evaluation on Education in Science and Engineering (REESE) program; and the Louis Stokes Alliances for Minority Participation (LSAMP) program. For example, the GK-12 program trains and then places graduate fellows in K-12 classrooms so they can bring their research expertise to students.

Outside of DEHR, other NSF programs have a substantial influence on teacher education. Katharine Covert, a program officer in NSF's Division of Chemistry, described some of the ways in which the research supported by the Directorate for Mathematical and Physical Sciences is integrated with efforts to support teacher learning.

The Division of Chemistry has about 1,500 research grants active at any given time, with an annual budget of about $200 million. A recent inventory of these grants showed that their educational components fall into a number of discrete categories: teacher workshops, research experiences for teachers, student workshops, laboratory experiences for students, providing help for science fair projects, judging science fairs, classroom visits or demonstrations, mentoring and training teachers and students, tours of labs, working with science museums, science camps, and curriculum and module development.

Not surprisingly, said Covert, a particular area of expertise within the Division of Chemistry is providing teachers and students with research experiences and visits to laboratories. A comparison of outreach activities sponsored by the Division of Chemistry and the Division of Elementary, Secondary, and Informal Education showed that the former engaged in more classroom visits, research experiences for teachers and students, and mentoring and training, while the latter specialized more in curriculum development and teacher workshops (Figure 4.1).

Evaluation of outcomes is an "Achilles' heel" within the Division of Chemistry, said Covert, partly because the programs are so variable. "These are community-driven programs, and even within the teacher workshops each is unique." Awards from the division go to chemistry researchers who "are not necessarily conversant with modern educational evaluation tools." There are exceptions of course, and the division strongly encourages ongoing evaluation. However, most of the evaluations take the form of questionnaires that are focused on the experience rather than outcomes. "The idea of a more scholarly evaluation that looks at the classroom impact is daunting," said Covert. "We proceed with a lot of goodwill, a lot of energy, a lot of wonderful anecdotes, and not a lot of hard outcomes."

NSF evaluates all research proposals based on their intellectual merit and their "broader impacts" on important societal goals. The conversation within the foundation, spurred in part by congressional attention to the issue, is transitioning from "what are these impacts" to "how do we track and assess them." NSF has set as a goal to deepen the understanding of these so-called broader impacts and to study the effectiveness of research activities in achieving them. This can be difficult to do with a large collection of relatively small projects, but each part of NSF is being challenged to evaluate the full range of outcomes of its activities.

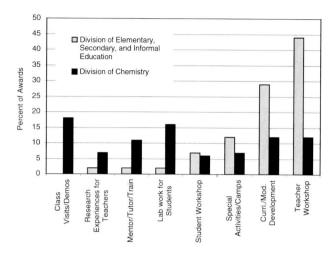

FIGURE 4.1 An estimated comparison of NSF outreach activities conducted in July 2006, shows that the Division of Chemistry conducts more outreach programs focused on research experiences and direct mentoring, while the Division of Elementary, Secondary, and Informal Education devotes more attention to workshops and special activities for students and teachers. SOURCE: Covert, K. 2008. K-12 Outreach Supported by NSF Chemistry. Presentation to Chemical Sciences Roundtable, Washington, DC, August 4, 2008.

Partnerships will be an important way to carry out evaluations of programs within STEM departments, Prival pointed out in the question-and-answer session. New activities can be added to existing evaluation programs, and discipline-based programs can partner with researchers in other academic departments who are skilled in evaluations.

The Division of Chemistry makes grants only to faculty members in STEM departments at colleges and universities, not to K-12 teachers or to K-12 schools or districts, Prival pointed out in response to another question. Yet faculty members often use part of their grants to work with teachers or schools, in part to achieve the broader impacts sought by NSF. Furthermore, faculty members often are interested in evaluating the impact of these activities to include in their reports to NSF.

THE DEPARTMENT OF ENERGY: CREATING TEACHER SCIENTISTS

An important part of the Department of Energy's (DOE's) work in education is directed toward K-12 teachers, especially toward middle school and high school teachers of science. "A lot of our emphasis is on educators, because we see this as giving us maximal leverage in the K-12 arena," said Jeffery Dilks of DOE's Office of Workforce Development for Teachers and Scientists. "We reach one teacher, and one teacher reaches many students, rather than addressing one student at a time."

A recent report from Dilks' office established a national goal of enhancing the ability of educators and the nation's educational systems to teach science and mathematics.[1] Priorities are to (1) enhance the capability of K-16 mathematics and science educators to boost student achievement in science and mathematics and provide a rich learning experience for students; (2) expand participation of women and underrepresented groups in the science-driven U.S. innovation system; (3) develop and support programs for students who wish to pursue science and technology careers at every step of the learning process; and (4) identify the appropriate roles and responsibilities for federal science and technology mission agencies in STEM workforce development, communicate these roles to all stakeholders, and serve as a catalyst for their involvement.

In particular, DOE's expertise is in experiential learning, Dilks said. "Whether it's undergraduates or teachers or graduate students, we take them out of their classroom setting and we place them into a laboratory and make them part of a research team to learn what research is all about." The goal is for teachers to take this experience back to their classrooms and translate it into something meaningful to their students, whether motivational, curricular, or pedagogical. DOE does not mandate how teachers should use that experience. Rather, DOE seeks to give them the tools to change what happens in the classroom. "We want to change the way teachers think about science so they can change the way that their students think about science."

DOE runs both pre-service and in-service programs for teachers. The largest in-service program is the DOE-Academies Creating Teacher Scientists (ACTS) program. It requires a three-year commitment from the teacher and is housed at a DOE laboratory. It is a residential program where the teacher lives at a laboratory for four to eight weeks, with a housing allowance, travel expenses, and a stipend. The teachers are integrated into a research team. Also, using an electronic portfolio, they create a professional development plan that outlines what they wish to accomplish over their three years in the program. The portfolio enables them to report back to program organizers about the changes they have made in the classroom.

The teachers take a content knowledge self-awareness survey before their first laboratory experience, at the end of their first year, and again at the end of all three years. In particular content areas, the teacher rank themselves on a five-point scale, from very knowledgeable to not knowledgeable at all. For example, they might be asked how much they know about the fundamental structure of atoms and molecules. They also take a professional practices inventory

before the first year to establish a baseline against which to measure changes.

The scores on these assessments do not necessarily lend themselves to evaluations of the program, Dilks said. For example, it may be more important for a teacher to realize over the course of a summer that he or she scores a 3 on a scale rather than a 4. "The core of our assessment process is to get teachers to be reflective and to think about what they do."

DOE puts about 150 teachers a year through 10 to 12 of its laboratories—an average of 15 teachers per laboratory—and the program could perhaps be expanded to 20 teachers per year per laboratory. However, there are hundreds of federal laboratories that are associated with the missions of federal agencies, many of them located in places that could reach out to more teachers than can the DOE labs. If each of these laboratories were able to take on 15 teachers, "you're starting to talk some real numbers," said Dilks. In addition, industry could run similar programs in its labs. "This is a scalable model if people cooperate."

Unlike most federal agencies, DOE has statutory authority to accept private funds to support its educational efforts, though the authority has been used very little so far. DOE therefore can partner with private industry to develop cooperative ventures in delivering outreach to teachers.

In the question-and-answer period, Dilks was asked how DOE reconciles its goal of attracting more females into science with the federal mandate to be inclusive in its educational programs. Dilks replied that the goal is to create environments that are receptive to all so that programs provide an encouraging atmosphere for young women.

Dilks also stated that DOE is seeking to develop an evaluation that would measure changes in student understanding that result from its programs. One complication in any such assessment is that there are many layers between the program and measures of student performance, he pointed out. Another complication is that student scores on many standardized tests are difficult to access directly, making it difficult to connect a test score with the practices of a particular teacher.

Also during the question-and-answer session, Kaye Storm of Stanford University recounted the experience of a teacher fellowship program that has been operating at Columbia University for many years. New York State has data available for the exam pass rates for students of teachers who have been through this program versus a comparison group. Although the sample size is small, the scores are higher for teachers who have gone through the program.

THE DEPARTMENT OF ENERGY: CREATING TEACHER LEADERS

Brookhaven National Laboratory, one of 17 national laboratories run by the Department of Energy, has approximately

---

[1]Office of Workforce Development for Teachers and Scientists. 2007. *Future Workforce Development*. Washington, DC: Office of Science, Department of Energy.

2,700 employees on a 5,000-acre campus on Long Island. It also provides approximately 4,000 guest researchers each year with access to instruments that no other institution can provide, including the Relativistic Heavy Ion Collider, the National Synchrotron Light Source, and the Center for Functional Nanomaterials.

Brookhaven also participates in the ACTS program, said Ken White, the head of the laboratory's Office of Educational Opportunity. One thing lab members often heard from teachers in surrounding areas is that they had no place outside their schools to do science. "They don't have places to do chemistry or professional development for themselves," White said. In response, Brookhaven has developed several ways for teachers and their students to be engaged in useful and authentic research activities. "We're actually getting the students out in the field and doing things that people care about."

One program takes advantage of the laboratory's unique setting. Water on Long Island comes from an aquifer that needs to be protected. In cooperation with teachers, the laboratory put together one-week summer workshops for 20 to 25 teachers focused on open space stewardship. The teachers learn how to do species classification, water sampling, and environmental testing. Then they team up with employees from the Suffolk County Department of Parks, the town of Brookhaven, or the U.S. Fish and Wildlife Agency to do sampling and analysis of water at sites close to their schools. The experts work with the teachers on the assays, tell them about the property, and accompany the teacher and students on field trips. Classes from elementary school to high school have taken data on public property and have shared their results with the agencies that have responsibilities for the property. It's a "community partnership," said White. The program promotes durable relationships, facilitates transfer of activities to classrooms, teaches environmental research skills, pairs teachers with properties, and engages students in authentic research. Money has come from DOE, from the schools, from other government agencies, and from grants that the teachers have written. A final celebration, conducted as a scientific meeting, enables students to present their research and celebrate their achievements.

A particular emphasis of the Brookhaven program is on creating teacher leaders. By the third year of involvement with the program, many teachers feel they are ready to use their new knowledge to take positive action at their schools. The laboratory is very interested in teaching the skills of science, including critical thinking, observation, analytical work, and gathering and analyzing data. By exposing teachers to a model of inquiry-based learning, the program prepares them to take such skills back to their schools.

Another Brookhaven project, done in partnership with Stony Brook University, takes place through the Center for Environmental Molecular Science. A two-and-a-half-day workshop for high school teachers enables them to remove metals from soil with citric acid. (Developing the workshop also led to the construction of commercial kits for the extraction of metals that are available through a private company.) "We send teachers away with the tools for how to do this and go back and be able to do chemistry in the classroom." Some teachers and their classes also have used the National Synchrotron Light Source to analyze contaminants in environmental samples.

During the question-and-answer session, the speaker and several members of the audience discussed the challenges in evaluating the effectiveness of these programs. The programs White runs are small and done on tight budgets, and finding money for a thorough evaluation, beyond simple questionnaires, is difficult. It also can be difficult to gather longitudinal data to track students or teachers over time.

One possibility is to rely on a single evaluation design that more than one grantee of a funding agency can use. If such an evaluation were available as a standard design and the same definition of outcomes were used, costs could be reduced appreciably below a stand-alone evaluation. Joan Prival of NSF stated that some of the foundation's evaluations are done that way, with a common set of data being collected from major programs. Individual projects also can adapt evaluation instruments developed for other projects, which cuts down on costs. Another possibility is for a funder to provide support for an evaluation center that will consult with programs about the design of an effective evaluation. However, because many different models exist, the same formula cannot be applied to every program.

5

Exemplary Programs

Major Points in Chapter 5

The ChemEd conferences held in odd-numbered years provide high school chemistry teachers with hands-on activities and other professional development opportunities.

The University of Pennsylvania Science Teacher Institute offers a master's of chemistry education for high school teachers that has yielded significant increases in chemical content knowledge for participants.

The AirUCI Summer Workshop for Teachers uses issues being studied in the Environmental Molecular Science Institute at the University of California, Irvine, to immerse teachers of chemistry and other subjects in scientific concepts and lab activities.

Evaluations of the "Terrific Science: Empower Teachers Through Innovation" program at Miami University in Ohio, which has provided more than 22,000 teachers with inquiry-based science workshops, demonstrate that the program has had a substantial influence on classroom activities and student learning.

Many outreach programs have sought to improve the quality of high school chemistry teaching in the United States. Presenters at the workshop described four such programs in detail. The programs were not necessarily chosen to represent the best of all the programs that have been offered, but they demonstrate some of the ways in which chemistry instruction can be dramatically improved.

THE CHEMED CONFERENCES

Since attending a chemistry education summer workshop sponsored by the National Science Foundation (NSF) at Michigan State University in 1965, Irwin Talesnick from Queens University has delivered somewhere between 1,500 and 2,000 presentations at professional development sessions around the world. Such an outcome probably would not appear in a program evaluation, yet that 1965 workshop was the "defining moment" of his life, Talesnick said.

The ChemEd conferences originated not long after. Following the 1972 Biennial Conference on Chemical Education (BCCE), a group of high school teachers and chemistry professors decided to organize a similar conference directed primarily toward high school teachers rather than college and university faculty. Since 1973, the ChemEd conferences have been held in odd-numbered years, while the BCCE has been held in even-numbered years. The ChemEd conferences attract 800 to 1,000 attendees, with about 80 percent high school teachers and 20 percent college and university faculty (percentages approximately reversed for the BCCE). The publication *Chem 13 News*, an informal magazine published by the Department of Chemistry at the University of Waterloo, has helped build support for the ChemEd conferences.

Many teachers pay their own way to the ChemEd conferences because of the difficulty of gaining support for travel and attendance. To encourage teachers to attend, ChemEd organizers build in a family program, with child care, a science camp for children, and various family activities. "The families get a vacation out of it, which makes it easier for the chemist in the family to travel to a different area every two years, enjoy the chemistry, and enjoy whatever else there is to be enjoyed."

The conference generally consists of four days of sessions, 50 percent of which involve hands-on activities, that encompass everything from 15-minute presentations to full-day sessions. Approximately one-third of the attendees at any conference have come to previous conferences, which is a measure of their success, said Talesnick. "Teachers have had only, in my experience, positive comments to make about the conferences." Furthermore, teachers forge friendships and collaborations at the conference that they maintain for years even if they are in widely separated locations.

However, most chemistry teachers say that they cannot attend the ChemEd conferences because of the expense. Talesnick therefore has been seeking financial support for the conferences to reduce the registration fee and associated costs. "If we had support from governments, industry, and so on—some of which we get but not enough—the registration fees could be reduced, the number of people will rise, and the costs will decrease." His other ambition is to make the conferences truly international, with attendance by chemistry teachers around the world. Achieving those two goals would have "a payoff for chemistry teachers, for universities, and for our students."

THE UNIVERSITY OF PENNSYLVANIA SCIENCE TEACHER INSTITUTE

The *Rising Above the Gathering Storm*[1] report cited the University of Pennsylvania Science Teacher Institute as a model program for in-service teacher preparation. "That's a great honor," said program director Constance Blasie, but "it's also a huge responsibility to provide excellent programming."

The program is based on the hypothesis that increasing the content knowledge of science teachers and influencing their classroom practices will increase the content knowledge and change the attitudes of the students they teach. "It's this hypothesis that drives our institute, drives our programs, and also drives evaluation," Blasie said.

The program was developed by University of Pennsylvania chemist Hai-Lung Dai and has been funded by the National Science Foundation since 2004, with additional support from the Rohm & Haas Company, the Camille & Henry Dreyfus Foundation, and the university. It is a collaborative effort of the School of Arts and Sciences and the Graduate School of Education. It offers two degree programs—a master's of integrated science education for middle school teachers and a master's of chemistry education for high school teachers. The program also offers a Science Education Resource Center that is supplied with many items that teachers can use while in the program or borrow to take back to their classrooms. In addition, the program provides mini-grants for which teachers can apply and two-day professional development workshops that have been co-developed and are co-presented by one of the teacher graduates and a University of Pennsylvania chemist.

The master's of chemistry education program began in the year 2000, so the ninth cohort of teachers began the program in fall 2008. To foster support for teachers within their schools and school districts, the program seeks to have each teacher attend with an administrative partner. As both partners learn about inquiry-based science, the administrators also learn what teachers need to make changes in their classrooms.

The program tries not to take teachers unless they have had more than two years of experience, so that they know how to manage a classroom and have decided that they want to remain in teaching. At the same time, many teachers in the program, who are drawn largely from the Philadelphia school district, are poorly prepared in chemistry.

Teachers take ten courses to earn a degree, eight on chemistry content and two on chemistry education. The program covers 26 months of coursework over three consecutive summers and two academic years, with the pedagogy courses delivered during the school year. The content courses, which are taught by chemists at the university, are organized not around lectures but around inquiry-based learning experiences. The courses also cover such topics as the nature of science, equity for students, and enduring understandings. The program relies heavily on technology and emphasizes nontraditional assessments. "This is not a program for everyone," said Blasie. "Teachers have to be absolutely committed."

To gauge its effects, the program has instituted an extensive evaluation effort. Two research associates work on internal and formative evaluations so that the program can make on-the-fly, real-time adjustments if its goals are not

---

[1]National Research Council. 2007. *Rising Above the Gathering Storm: Energizing and Employing America for a Brighter Economic Future.* Washington, DC: The National Academies Press.

being achieved. As part of a broad external evaluation, the research associates also gather data on such topics as content knowledge and teacher understanding of the nature of science. Teachers take a specially designed chemistry content examination before they enter the program and again after they have completed all the coursework.

The content examination has revealed that teachers demonstrate a highly significant increase in chemistry content knowledge over the course of the program. They also develop a better understanding of the nature of science.

To assess changes in teaching practices, teachers prepare a baseline teaching portfolio at the beginning of the program that describes a four- to five-day lesson plan. At the end, they prepare another such lesson plan based on their thesis topic. Program evaluators then use a lesson plan analysis tool to analyze the two plans. The analysis shows that the later lesson plans reflect a much deeper understanding of how to deal with equity issues in the classroom, how to use technology, and how to encourage students to practice their own analytic skills. The one area in which they do not improve, Blasie noted, is in using formative assessments to understand what students know and what their misconceptions are.

Teacher and student questionnaires compare the characteristics of classrooms both before and after a teacher participates in the program. Results from both perspectives show significantly increased use of standards-based instruction once teachers have graduated from the program.

Measures of student performance have been hampered by the fact that different groups of students are being tested each year. However, a content examination given to successive groups of students showed that students of program graduates did significantly better than students of teachers before they entered the program. Also, student questionnaires revealed that students had a much better attitude about science after their teachers attended the program.

Much more can be done with evaluation data, Blasie noted. For example, the electronic portfolios that teachers keep could be probed for many different types of information. One interesting suggestion made during the question-and-answer period addressed the difficult issue of finding a control group against which to make comparisons. Eric Jakobsson from the University of Illinois discussed a project called Chemistry Literacy Through Computational Science. As a control, half of the teachers recruited to the program were delayed for a year and served as a control group for the teachers who began the program.

THE AIRUCI SUMMER WORKSHOP FOR TEACHERS

The AirUCI Summer Workshop for Teachers was founded in 2005 as an outreach program of the NSF-supported Environmental Molecular Science Institute, with additional support from the Camille & Henry Dreyfus Foundation. Since 2005, four workshops have been offered that have served about 20 teachers in the region annually. Most are from public high schools and middle schools located near the University of California, Irvine (UCI), and most teach chemistry at least part of the day. Some also teach environmental science, physics, earth sciences, biology, and integrated sciences. Most have bachelor's degrees, with a small number having Ph.D.s and a small number having no college degree at all. The workshop lasts for two weeks and teachers are paid a stipend of $1,000, which is less than they would get for teaching summer school. "We don't have people who are in it for the money," said UCI's Sergey Nizkorodov. The program estimates that each teacher interacts with approximately 150 students per year. The program therefore is able to reach 3,000 additional students each year, along with the students' parents and members of the community.

The hypothesis behind the program, said Nizkorodov, is that "if we convey enough excitement to the teachers, they'll become better teachers and affect students that way." The workshops involve faculty, graduate students, undergraduates, and doctoral researchers—"everyone who participates in the AirUCI Institute." Prominent faculty at UCI deliver lectures on a wide variety of topics, including atmospheric chemistry, climate change, air pollution, the interaction of life and matter, surface science, and hydrogen bonds, and guest lecturers who are working at the institute provide talks on additional topics.

The workshop also features labs adapted from those that are offered to upper division undergraduate students, scaled down so they can be completed in four hours. The labs use equipment recently purchased and refurbished through a grant from the Camille & Henry Dreyfus Foundation. Groups of three or four teachers work with a graduate student from the institute, with the graduate students receiving $1,000 for their assistance. For example, one lab uses spectrometry to measure the amount of alcohol in vodka; another measures the concentrations of polycyclic aromatic hydrocarbons in cigarette smoke; another measures the particle removal of auto emissions by air purifiers; and another measures aromatic compounds in gasoline. A newly developed lab uses laser-induced breakdown spectroscopy to analyze metals. Besides five wet labs in each workshop, two computer labs are offered—one based on a model of air pollution in the Los Angeles basin and the other based on the greenhouse properties of various pollutants. Finally, at the end of the program, the teachers do a half-day lab tour of institute activities.

"We don't do a very good job of evaluating our program," said Nizkorodov. Mostly, the program has relied on self-evaluations by teachers immediately following the workshop. Recently, however, the program has been able to follow up with teachers in the previous workshops with an anonymous survey. When asked the question, "Have you been able to integrate any new information from this program into your course syllabi?" 84 percent responded, "Yes, to a certain extent." An additional 13 percent responded, "My syllabi have changed significantly as a result of taking this course." When asked the question, "Do you feel you are in a better position to discuss topics associated with climate change, air pollution, and atmospheric chemistry with your students and colleagues after participating in this program? "97 percent responded, "Yes, my understanding of these topics definitely improved a lot."

Teachers have many opportunities to attend other workshops, Nizkorodov noted, though perhaps not with equipment as sophisticated as that available at the institute. The teachers at the workshops had attended an average of five to ten workshops before. The survey asked, If you attended more than one teacher development program over the last 10 years, please rate this program relative to the others. Thirty-two percent of the teachers felt that it was the best program they had participated in so far, while another 42 percent said it was superior to the other programs they had attended. When asked to rate the most effective aspects of the program, the majority of teachers cited the close interactions with faculty members, with the laboratory experience being the second-most cited factor.

The AirUCI Institute plans to continue the workshops for the foreseeable future, which may provide additional opportunities for evaluation. Nizkorodov also noted that the workshops provide a valuable opportunity for graduate students and postdoctoral researchers to learn to accept responsibility for training teachers and communicating with the public.

TERRIFIC SCIENCE, 25 YEARS OF OUTREACH IN CHEMICAL EDUCATION

Since "Terrific Science: Empowering Teachers Through Innovation" was founded 25 years ago at Miami University in Ohio, more than 22,000 teachers have participated in the program. The leaders of other programs often ask how Terrific Science has reached such a large number of teachers, said Gil Pacey, a professor of chemistry and biochemistry at Miami University. The answer is that all of the workshops offered through Terrific Science, which range in length from a few days to two weeks, offer credit; Miami University has waived all tuition in most cases; and funding agencies have helped pay for housing and have offered stipends to teachers. "We hand out quite a lot of carrots," Pacey said.

Terrific Science, a nonprofit organization run by Miami University's Center for Chemistry Education, has produced more than 250 professional development programs; more than 80 books, kits, and other resources; and an online repository of more than 200 resources for teachers (*http://www.terrificscience.org*). The program has received more than $16 million in federal, state, and private funding to increase scientific literacy and to stimulate interest in and understanding of science.

The vision of the program is to create engaging, motivating, and fun learning experiences. "We bring chemistry and the companion sciences to life for teachers and students at all levels," said Pacey. Workshops are organized around hands-on activities, so that instructors do things with teachers and not for them. Teachers learn via modeling and constructive discourse and are encouraged to take risks in a supportive environment. In turn, teachers are encouraged and supported to take activities back to their schools and use them with their students. Students experience the fun and excitement of doing inquiry-based science rather than having science done to them. By nurturing students' curiosity, science motivates them and inspires their innovation and creativity. Doing science this way also promotes critical thinking and problem solving, which is "absolutely necessary" in today's economy, Pacey said.

The program partners with approximately 150 colleges and universities, 1,000 school districts across the United States and abroad, and 100 other organizations. For example, the South Korea Metropolitan School district recently sent 50 people for two weeks in two consecutive summers to participate in the program. Corporate partners also have used the program as a conduit to provide nearby schools with desperately needed supplies.

The Center for Chemistry Education has established a set of best practices that call for the extensive use of teacher leaders, mentoring teams, and collaboration with stakeholders, including government and industrial labs. The best practices also call for learning activities that are content rich, pedagogically strong, and extended over time. Teachers and administrators participate in curriculum development, implementation, and evaluation using what they have learned in workshops. For example, after learning to measure pollutant levels in lake water, participants in a workshop might be asked what kind of inquiry-based module they could develop for their students, given the constraints on equipment, supplies, and other resources. Teachers then implement the module in their classes, test it, improve it, and disseminate it to other teachers.

The program follows up with teachers for at least a year after each workshop. For example, teachers might meet with Terrific Science educators to discuss the implementation of a newly developed module. Some graduates of the program also become facilitators for other teachers and eventually teacher leaders who run workshops themselves. Pacey estimated that 10 percent of the teachers who go through the program give papers at regional and national meetings based on what they have accomplished. He also estimated that the average graduate of the program reaches 35 other teachers in the first two years after the workshop, greatly multiplying the program's effects.

Pacey cited a number of lessons learned from the program. Scientific explanation without related experience has little impact on learners. Lifelong scientific literacy begins with the attitudes and values established in childhood. How physical science is taught is as important as what is taught.

Instruction should build on children's innate curiosity, provide firsthand experiences that involve all of the senses, be connected to everyday experiences and observable phenomena, and provide connections among ideas.

Evaluations of the program have shown that the students of participants spend more time doing laboratories that involve taking measurements and doing graphical analyses of data. The students of teachers reached by the program also spend more time testing student-generated hypotheses. In tests of physical science learning supported by the Ohio Board of Regents, post-test scores for students in grades 3 through 9 increased dramatically when their teachers had gone through these programs. "Teachers learned how to translate information to their students in a more effective way," Pacey said.

Teachers' comments about the program are extremely positive, as are comments from their students. In particular, students express more interest in science-related careers after their teachers have participated in the program.

The program also has found it necessary to do outreach to parents to convince them that science education is important for their children. But "we have a major public relations problem, probably across the whole country," Pacey said. Ohio offers many examples of good and available jobs that are related to science and technology. For example, Wright-Patterson Air Force Base in Dayton will need 7,000 bachelor's, master's, and Ph.D. scientists and engineers to replace retiring workers in the next five years. "We don't know where we're going to get them, so we have to do a sales job on parents," said Pacey. "We probably also have to do a sales job on [high school] counselors."

6

Activities by Nonprofit and For-Profit Organizations

Major Points in Chapter 6

The Bayer Corporation's Making Science Make Sense program supports outreach to teachers and other activities designed to foster a well-educated workforce and a scientifically literate public.

The Achieving Student Success Through Excellence in Teaching program provides high-quality teaching materials and professional development for elementary school and middle school teachers throughout Pennsylvania.

The American Chemical Society sponsors several programs for high school chemistry teachers and can have a substantial influence on high school chemistry through its affiliates, local sections, and clubs.

The Hach Scientific Foundation offers scholarships for chemistry majors and for people working in chemistry-related fields who intend to become chemistry teachers, and it supports training programs for current teachers.

The grants program of the Howard Hughes Medical Institute emphasizes evaluations of the education programs it supports and wide dissemination of the results of assessments and of information about successful programs.

Surveys and interviews conducted with representatives of 37 foundations active in science education found that foundations support a broad range of activities but, with several important exceptions, have gathered few data about the effectiveness of the programs they fund.

THE BAYER CORPORATION

Since the 1960s, scientists, engineers, and other employees of the Bayer Corporation have been volunteering in local schools to enhance science education. Most of these efforts were ad hoc, said Bayer's Bridget McCourt, until 1995, when all of the company's educational efforts were brought together to form the Making Science Make Sense program. "It is our premier corporate social responsibility program

here in the United States," said McCourt. It is designed "to help teachers teach and to help students learn the way that scientists do."

The program, which has been recognized by several national awards, extends from preschool to graduate school and beyond and addresses two populations of students. First, it seeks to prepare a well-educated workforce that can fill high-level jobs, including jobs in scientific and engineering research. Second, it is directed toward producing a scientifically literate public so that Bayer can be a successful business. "When we're either establishing a new site in the community or having a crisis in the community, we need [the public] to be based in science. . . . We need them to have a basic understanding of chemistry, biology, physics, and other science fields so that they're . . . educated voter[s] and neighbor[s]."

Science education is particularly well suited to fostering the kinds of skill people need in today's world, said McCourt. It teaches creative thinking, critical thinking, team building, and adapting to change. Science education "is not just about educating the next generation of scientists, engineers, and mathematicians. It's about equipping students with the skills that they'll need in whatever job they go into."

Making Science Make Sense has three components. The first is systematic science education reform. Schools need help to move beyond traditional teaching approaches toward inquiry-based learning. For that reason, Bayer projects support the national science education standards and incorporate substantial professional development for teachers. For many classrooms, McCourt observed, inquiry-based learning involves a "complete shift" in the way teachers are teaching and the way students are learning.

The second component of the program is public education and outreach. Led by former astronaut Mae Jemison, this component of the program has featured a variety of partnerships on both the national and the local levels. Through these partnerships, Bayer has been able to learn where the company's efforts are needed and how those efforts can help. For example, a partnership with the American Chemical Society has led to efforts to address teacher development and diversity in science, technology, engineering, and mathematics (STEM) education. This partnership resulted in a recent set of three-day workshops on green chemistry for high school teachers. It also created new internships in Texas that give disadvantaged students an opportunity to experience chemistry careers through hands-on summer internships. In addition, a partnership with the Carnegie Science Center in Pittsburgh has led to a program in which high school students are trained and employed to do scientific experiments and field trips both during the school year and during the summer with small groups of elementary students. Since its inception in 2000, this program has been reaching about 250 elementary school students per week during the school year and nearly 1,000 students on average in the summer, and each of the high school students who has participated in the program has gone on to college, often as the first person from his or her family to do so.

The third component of the program is based on employee volunteerism. As has been the case throughout Bayer's involvement with education, volunteers continue to work with individual students, teachers, and schools to bring meaningful and enjoyable scientific experiences into the classroom. Together, the three components of Making Science Make Sense "form a comprehensive and integrated program that is driving results and impacting lives."

As Bayer has expanded into new communities, it has brought the program to new places. Branches of Bayer outside the United States also have been instituting versions of the program in their home countries, including Japan, Colombia, Italy, India, and the United Kingdom. In addition, Bayer has encouraged the involvement of other corporations in educational initiatives, in part by sponsoring forums on science education. A 2006 forum on educational diversity held in Washington, DC, for example, attracted more than 150 STEM industry and organization representatives, federal and state government officials, and others in the nonprofit and science education fields.

As with many of the other programs discussed at the workshop, assessment of the Making Science Make Sense program is difficult, McCourt acknowledged. The company requires projects supported through the program to perform assessments and provide the company with progress reports. Bayer also has supported efforts to assess the state of science education in the United States, including an annual survey on STEM programs, policies, and practices.

"All of corporate America has a role to play in improving science education and science literacy across the country," said McCourt. "We believe that Making Science Make Sense is an effective program for our corporation, and we are committed to continuing the program in the years to come."

During the question-and-answer session, Ken White from Brookhaven pointed out that if studies demonstrate the value of programs such as Making Science Make Sense, other companies might be influenced to initiate and participate in outreach efforts. McCourt responded that the initial success of a program can drive future successes, especially when volunteers come back into the workplace and describe their accomplishments to others. However, she also noted that one challenge she faces is to maintain the continuity of the volunteer effort over time. Half of the Bayer workforce has joined the company just in the past five years. "I have to continually reorient people to the program, introduce them to it, explain to them what it is," she said. A great advantage at Bayer is that the leadership of the company supports the program and encourages employees to participate. "It's not seen as a detraction from their position but rather as an addition to their role in the company."

ASSET: ACHIEVING STUDENT SUCCESS THROUGH EXCELLENCE IN TEACHING

One program that the Bayer Corporation has supported, along with other funders, is Achieving Student Success Through Excellence in Teaching (ASSET). It was created in Pennsylvania in 1994 as an independent, nonprofit, educational reform initiative dedicated to continuously improving the abilities of teachers and the performance of students. Its vision is to be a leader in developing and implementing effective, innovative programs, products, and practices that align teaching to learning. It is focused on kindergarten through eighth grade, which is essential to establish "a strong foundation for you at the high school level," said its executive director Reeny Davison. ASSET has become "the leading science education organization in classrooms throughout Pennsylvania."

The hypothesis behind the program is that high-quality materials and high-quality professional development will produce more effective teachers and better-performing students. It has employed standards-based curriculum materials, centralized materials support, assessment, and involvement of the administration and communities to create a national model for effective science education reform. The program has drawn heavily on materials and methods developed by the National Science Resources Center, which is a joint project of the Smithsonian Institution and the National Academies. The program also uses materials from other sources, such as the Full Option Science System from the Lawrence Hall of Science. "As a nonprofit, we are free to become what teachers need us to be. If teachers don't need us, we will go out of business."

ASSET's Materials Support Center purchases standards-based materials and stores, cleans, refurbishes, and distributes those materials, in some cases with hands-on assistance from its corporate sponsors. Schools choose the kits they want to use, which range across the earth, life, and physical sciences as well as technology and engineering. "Like a good business, we give our districts choices," said Davison. "We don't tell them what they have to order. They order what's right for their curriculum and for their teachers."

ASSET also supports professional development in the form of teachers' teaching teachers. "When you have another teacher standing in front of you, there is instant credibility, because they can say that when I did this I found that this trick helped."

In 2001 the program transitioned to a fee-for-service organization, which required that it continually develop new products and services for teachers, in part through partnerships with private organizations. In 2006 the State of Pennsylvania launched the "It's Elementary" initiative and arranged with ASSET to expand its program throughout the state. "No one in the Pennsylvania Department of Education or the Governor's Office designed the program," said Davison. "We got to design it, coordinate it, and implement it according to the things that we have learned in the last 10 years." ASSET would like to become a professional development center that teachers can rely on in a standards-based environment.

ASSET is currently serving 164 school districts, 6,392 teachers, and slightly more than 180,000 students. Teachers engage in multiday workshops over more than one year, usually focusing on one curriculum module each year.

The program has contracted with Horizon Research, Inc., to do evaluation research, including comparing student scores with the amount of professional development teachers have undertaken. The results show that students whose teachers participated in three days of professional development scored significantly higher than students of teachers who participated in two days or less. Furthermore, student achievement was greater the second time the teachers implemented a module to which they had been exposed during an ASSET workshop.

Davison called for cooperation among programs to address the full range of problems facing teachers and students. "There can't be too many of us," she said. "The time for competition is over. It is all about collaboration."

THE AMERICAN CHEMICAL SOCIETY

Mary Kirchhoff of the American Chemical Society (ACS) briefly described some of the activities undertaken by ACS to improve high school chemistry education. ACS conducts a number of summer workshops, including a three-day residential workshop on green chemistry (partially sponsored by the Bayer Corporation) and a five-day workshop on bringing chemistry into the community. "One of the things that struck me throughout the workshops is how much the teachers appreciate the opportunity to talk with each other," Kirchhoff said. Teachers from different kinds of schools were able to describe both the particular challenges they faced and the issues common to all teachers.

Other ACS activities provide training for teachers of advanced placement (AP) and international baccalaureate courses in chemistry and offer workshops for middle school science teachers and their supervisors. A new edition of the book *Chemistry in the National Science Education Standards* addresses standards and provides models for meaningful learning in high school chemistry classrooms.[1]

The ACS has looked periodically at the idea of forming a stand-alone high school chemistry teachers association. Although the idea has not gained traction in the past, said Kirchhoff, she planned to bring it up again with the society's Committee on Education. "Out of 160,000 members of the ACS, only a couple of thousand are high school chemistry teachers. Clearly, they are not finding the value that we could be providing to them."

ACS has large networks of members, local sections, student affiliates, and high school chemistry clubs, all of which can have an influence on high school chemistry education. Where resources are not available in a particular school or

---

[1]S. L. Bretz, ed. 2007. *Chemistry in the National Science Education Standards,* Second edition. Washington, DC: American Chemical Society.

district, the ACS can step in and provide a service directly or foster a partnership that could meet the needs that exist.

The ACS also has been working with organizations in higher education such as the National Association of State Universities and Land Grant Colleges to improve chemistry education, including the education of undergraduates who become chemistry teachers.

THE HACH SCIENTIFIC FOUNDATION[2]

Clifford Hach was a chemist who worked on the Manhattan Project in the 1940s and started the Hach Company, which was an analysis, instrumentation, and water chemistry firm. Located originally in a one-room building in Ames, Iowa, the company grew and moved to Colorado in the 1970s. In the early 1980s, it formed the Hach Scientific Foundation to provide scholarships to future chemists. The foundation became fully funded when Hach died and the company was sold.

Several years ago, Clifford Hach's grandson Bryce Hach, who is executive director of the foundation, decided to drive to each of the scholarship recipients and ask them why they chose to study chemistry. "I was a biology major myself, and I wasn't the greatest chemistry student in the world. I thought chemistry was really hard, so I was curious," Hach said. "At least 90 percent of them said that the number one influence that led them into chemistry was a really good high school chemistry teacher." That led Hach to give greater consideration to the importance of these teachers. Only about a quarter of high school chemistry teachers have a degree in chemistry, Hach said, and less than half of them minor in the subject. Motivated by these observations, the foundation decided to broaden its involvement in chemistry education.

In the 2007-2008 academic year, the foundation began offering scholarships to chemistry majors who plan to go into teaching. At least two $6,000 scholarships are provided at each of the land grant universities in the country, which ensures that the program will have a national reach. The scholarships are available for undergraduates at any level, including undergraduates who want to spend extra time in a university to take education classes. The foundation wants to reach students who are thinking about going into research, industry, the pharmaceutical industry, medical school, or other destinations and get them thinking about teaching. It wants to "create teachers where there otherwise weren't any."

In addition, the foundation has created a second-career chemistry teacher scholarship program for people who have worked in a chemistry-related field and are looking to go into teaching. This $6,000 scholarship can be used at any college or university in the country as long as the student has been accepted into a program to work toward a master's in education. A $3,000 scholarship is offered for part-time students who continue to work or support a family. "We've had scholars ranging from their early 20s to their early 60s," said Hach.

With just three full-time staff members, the foundation provides universities with the criteria for the chemistry major scholarship, and the universities administer the scholarships, usually through the chemistry department. The foundation chooses the second-career scholars itself, with advertisements in chemistry publications to inform potential recipients about the program.

The foundation also has decided to provide in-service support to chemistry teachers, so it has begun a program to offer a $1,500 grant to any chemistry teacher who would like to improve his or her teaching skills. A one-page application on the foundation's Web site (*http://www.hachscientificfoundation.org/home.shtml*) asks how teachers are going to use the funds and how the impact of the funding will be measured. "We want to make the application process as simple as we can," said Hach. Later, teachers write a one-paragraph summary of how the grant was used that is posted, by state, on the Web site.

With very little advertising, the foundation received more than 200 applications in the first two-and-a-half months of the program and was able to grant 178 requests in more than 40 states. The program "was far more exciting and far more involved than we ever thought it would be, and we're really glad to continue the program. Certainly it shows that there's a lot of untapped potential here."

In northern Colorado, the foundation runs a program to bring together almost all of the chemistry teachers in four local school districts to engage in workshops organized around the Process Oriented Guided Inquiry Learning, or POGIL, approach. The program begins with a three-day workshop, followed by subsequent one-day and two-day workshops. The foundation pays for substitutes while teachers attend the workshops. POGIL "transforms the chemistry class from a passive learning environment to an active one," said Hach. "Students have to teach each other. They work in small groups. They're doing real research. They have to take the onus of education on themselves."

Two chemistry education Ph.D. students are doing their dissertations on the impacts of these workshops on learning. Initial assessments have demonstrated a 15 to 20 percent increase in the grades of students whose teachers have participated in workshops and a 15 to 20 percent increase in students' going on to higher levels of chemistry. "The results from this program will be available on our Web site as soon as they're formally released," said Hach. "Everything is going to be transparent to the public."

---

[2]In January 2009, the Hach Foundation announced that it plans to transfer the foundation's funds and assets to the American Chemical Society (ACS) to administer the grants described in this section. For more information see: Raber, L. 2009. Philanthropy: ACS Receives Hach Funds. *Chemical and Engineering News* 87(4):7.

THE HOWARD HUGHES MEDICAL INSTITUTE

The Howard Hughes Medical Institute (HHMI) is primarily a biomedical research organization, funding more than 300 scientists and their associates in research laboratories across the United States. However, HHMI also has a grants program that supports precollege science education, along with undergraduate and graduate education and research. In the most recent precollege competition in 2007, $22.5 million in grants were awarded over five years to 31 institutions to engage in educational outreach.

HHMI has learned a number of valuable lessons since the grants program was initiated in 1987, according to HHMI's Patricia Soochan. The first is to assume nothing and assess everything. Assessments should be done early, often, and comprehensively and should be quantitative as well as qualitative. "Assessment should be used to adjust the program as necessary and make sure the grantee is on the right track," said Soochan, "not . . . wait to the end to show the foundation that the grant has worked."

HHMI also emphasizes dissemination, both of the results of assessments and of successful programs. Networking with others helps to ensure that useful models are replicated and mistakes are avoided. Publishing the results of assessments helps to disseminate results widely.

From 1988 to 2008, HHMI's grants to undergraduate institutions totaled $767 million, and 22 percent of that amount—about $170 million—went to precollege and other outreach activities. Those grants have served about 85,000 teachers from preschool to high school in programs lasting more than two weeks, with many more served in shorter programs. The precollege programs are very heterogeneous, said Soochan. Most focus on biology, but some focus on chemistry, physics, and other areas of science. They range from 10-week summer research experiences to workshops that meet periodically during the school year. Among the features characterizing successful programs have been involving teachers in the early stages of program conception and development, treating teachers as partners, incorporating educational standards, using master teachers, providing continued resources such as undergraduate teaching assistants and equipment libraries, encouraging networking, providing subsequent experiences, and including support for evaluation.

Soochan described two examples. A grant to Emory University supported teams of middle and high school science teachers, graduate students, and undergraduates on a year-long project to create model inquiry-based curriculum materials that are aligned with the Georgia and national standards. From 2003 to 2007, teams that included 48 teachers implemented 166 new units in 150 classes of more than 4,000 students. Evaluation of the program included surveys of teachers, audits of lesson plans, measures of student performance, reviews of student portfolios, comparison of college entry rates, and focus groups. "In science education we have learned that an arsenal [of assessment strategies] is really what's needed. . . . Assessment has to be very creative, and you have to be willing to do many different types."

The other program she described was at Occidental College, which had the goal of improving high school biology and chemistry students' laboratory instruction by enhancing their teachers' knowledge and classroom application of modern instruments, techniques, and experiments. The program consisted of 13 experiments developed and tested in classrooms by a steering committee of about a dozen high school teachers and five college staff members. The experiments, which conformed to the science framework for California public schools, employed a biochemical focus to enhance and bridge the biology and chemistry curricula. Each teacher who participated in the program attended a two-week summer institute focused on the details of the experiments. Activities reflected the background of the experiments and instruments, hands-on practice with the experiments using both inquiry-based and traditional instructional models, and pedagogical discussions of how to incorporate the experiments into the curriculum at different levels.

The program also used high school students selected by their teachers from the previous year's classes and trained to operate the specialized instruments and equipment. The students then assisted in the classroom during the labs. Participating teachers generally used a specific experiment with three to five classes, with many teachers using it for all of their classes. From 1992 to 1995, teachers conducted more than 38,000 student experiments. The experiments also were adapted to a wide variety of other settings, ranging from AP classes to other science classes.

A statistical analysis of the responses on student questionnaires suggested a significant positive change in students' attitudes toward science and toward the equipment.[3] A survey of teaching assistants indicated that their involvement increased their interest in majoring in science as undergraduates and their interest in a science teaching career. Furthermore, survey results strongly suggested that teachers experienced significant growth in their knowledge of biology and chemistry concepts and the use and theory of the instrumentation underlying the experiments. The positive impact of the program on teacher content knowledge and classroom activities was strongly substantiated by the direct observations of the program's outside evaluator.

WHAT ARE FOUNDATIONS DOING?

Given that foundations support a wide variety of education reform efforts, Sandra Laursen and Heather Thiry at

---

[3] C. Craney, A. Mazzeo, and K. Lord. 1996. A high school-collegiate outreach program in chemistry and biology delivering modern technology in a mobile van. *Journal of Chemical Education* 73(7):646-650.

the University of Colorado at Boulder, with support from the Camille & Henry Dreyfus Foundation, set out to learn more about the outlooks and practices of foundations. Their approach was to ask foundations that support activities in secondary chemistry education a series of questions: What do you do? What evidence do you have about how it works? What do you conclude from the evidence? How does the evidence shape your practice?

First they analyzed the Web sites and available publications of 37 foundations identified as key players in science education. Then they conducted surveys and in-depth interviews with 16 selected foundations. They divided the activities supported by foundations into five broad categories (Figure 6.1). In the first category—direct support for students—they placed scholarships and competitions. Examples include competitions "that have students inventing things or solving problems," said Laursen, who summarized the study's findings at the workshop, or scholarships "for high school students to do summer research or to have some kind of extra learning experiences."

The second category—classroom support—includes programs directed at teachers or individual classrooms, such as grants for equipment, programs to develop curricula, or professional development for teachers. The third category—informal education—includes all activities beyond the K-12 educational system, such as support for museums, science centers, summer camps, and after-school programs.

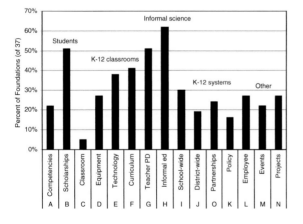

FIGURE 6.1 The percentage of foundations engaged in supporting secondary science education was highest for informal science and lowest for activities focused specifically on students in classrooms. SOURCE: Laursen, S., & Thiry, H. (2008, January). *What Do We Know about What Works? Review of US Foundations' Programs in Secondary Chemistry Education.* (Report to the Camille & Henry Dreyfus Foundation) Boulder, CO: University of Colorado at Boulder, Ethnography & Evaluation Research.

Support for K-12 systems, the fourth category, can go to schools, to districts, to partnerships, or for policy development and implementation. Also, a fifth miscellaneous category includes activities such as employee volunteerism, special events, and projects such as film or web projects.

The researchers attempted to attach dollar amounts to these activities, but the range of programs and activities made this impossible, especially given that education accounts for about 25 percent of all philanthropic giving. Nevertheless, by establishing these categories, the study sought to examine activities that the foundations deemed important. "Our idea was to look at these activities as a way of saying, What do people think works?"

From the broad analysis of 37 foundations, Laursen and Thiry discovered that corporate foundations tended to support different activities than private foundations. Corporations tended to fund student scholarships and competitions and small classroom grants to entrepreneurial teachers. They also tended to target their home communities. Private foundations were more likely to engage with districts, systems, or policy. Both supported teacher professional development, which they see as a high-leverage strategy.

Informal education was popular with both types of foundations, partly because it did not directly involve school systems. "You don't have to deal with all that bureaucracy, all those state standards, and all those rules," said Laursen. "They see K-12 systems as difficult, as too big a ship to turn." Informal education is also a way to inspire and motivate students and build their interest in science. It is difficult to measure the impacts of these activities, but "I think we all believe and have seen examples in our own lives about how that works."

Laursen and Thiry hoped that, in their interviews with foundation representatives, they would uncover stores of data about the effectiveness of programs that had not been analyzed. This turned out not to be the case. "They are busy. They are on the road. These people . . . are doing a lot of good things. [But] that mine of data doesn't for the most part exist."

On the contrary, the researchers found that fairly few data are collected and that the sources of information are mainly grantees' reports and site visits. Most of the information is about the populations served and the activities conducted, with uneven internal evaluation and little external evaluation. Most foundations know what happens to whom, but they know little about whether, how, or why it works. However, said Laursen, the researchers talked with very insightful program officers and found very interesting initiatives under way.

From these interviews, the researchers culled a number of "best practices" in grant making. These practices are "experienced people's advice, but not necessarily evidence-based advice," said Laursen. "They have gone out and have seen things and have watched things and have paid attention to similarities and differences. They don't necessarily have data in hand."

In setting directions, foundations should draw on the research literature, on national reports, and on observed trends. They should seek to have an impact through either breadth or depth. "They are making strategic choices. Do we spread our resources over a wider area and go for impact by having lots of people participate, or do we go for depth in a smaller area or a smaller targeted project?" Foundations also are evaluating their own work to set future directions. The data drive them in directions that they might not have considered before.

General elements of strong project design include building stakeholder support, beginning with a needs assessment, using the research literature, involving scientists and engineers, and addressing sustainability up front. "What happens when the foundation money ends?" Foundation officers were interested in seeing plans that had a longer-term vision of how to keep programs going once funding is gone.

Best practices in teacher professional development include aligning content with the curricula teachers are using in class, aligning with state and national science standards, strengthening teachers' content knowledge while linking to pedagogy, incorporating follow-up in professional development, providing time to reflect and network, and modeling and discussing effective teaching and learning methods. Evaluating teacher professional development is not straightforward, Laursen observed, partly because the desired effects are far downstream, but evaluation efforts are necessary.

Laursen and Thiry found several intriguing examples of foundations that were trying to improve their evaluation practices. Accountability ensures that foundations can learn from the activities they support. "As one foundation officer said, in the end the board is going to look at you and say, 'Well, what happened?'" Sometimes knowledge can be generalized from one program across a range of programs so that general principles can be distilled. Having some sense of the impact of a program can be motivating for funders and practitioners and can engage each in further activity. As Laursen pointed out, other researchers have speculated that the use of good evaluation could multiply the payoffs from foundation resources at least severalfold.

The Bill & Melinda Gates Foundation has established an entire evaluation office and has set up metrics for the schools it is supporting. This may not be a realistic strategy for smaller foundations, but these organizations may be able to use common and shared evaluation tools. For example, the Noyce Foundation is compiling the surveys, interviews, and other methods that are publicly available to study the impact on students from informal science education experiences. Once these instruments have been identified, gaps can be located and tools can be supplied to grantees for use in evaluating projects. In contrast, the Burroughs Wellcome Fund is developing the capacity of its grantees to evaluate their own work. The fund has an evaluation team that leads workshop, does consultations, and coaches grantees on how to identify goals, measure progress toward those goals, analyze data, and draw broad conclusions across projects. The evaluation work is supported by a tax of about 1 percent on each of the grantees. "Across all of their grants, this adds up to enough money to fund this kind of effort."

Laursen also cited a tool developed by her colleagues Elaine Seymour and Tim Weston called Student Assessment of Their Learning Goals (SALG). It is a publicly accessible assessment tool that faculty can use to ask students what they gained from a course and what aspects of a course helped them learn. It is online and free, with core questions and optional additions, at *http://www.salgsite.org*. The instrument has about 12,000 users so far who have customized their versions. A similar instrument, also available on the SALG Web site, is the Undergraduate Research Student Self-Assessment (URSSA), which is a research-based technique for assessing what students get from doing undergraduate research.

Chemists need to apply evidence-based methods in their education work as well as their science, Laursen concluded. They need better evidence about what works to shape the design and implementation of projects, to guide the choices of projects to fund, and to learn from their own and other's mistakes and successes. They need to think about their objectives and how to measure progress toward those objectives at the beginning of a project, not at the end. Funders and program developers alike have an interest in sharing processes and tools for evaluating the outcomes of educational outreach efforts.

In the question-and-answer session, Tom Keller of the National Academies' Board on Science Education noted that the National Science Foundation has just released a framework for informal science education. It is a good starting point for anyone interested in evaluating such programs, he said.

7

Future Actions

Major Actions Suggested by Workshop Participants

Coordination within and among organizations, including coordination between teachers and teacher educators, is essential to improve chemistry education.

Evaluations that commence at the beginning of chemistry education outreach programs, that examine failures as well as successes, and that are widely distributed are important for developing a knowledge base on which future programs can build.

The development of a more effective teaching corps requires both ongoing professional development and careful attention to the recruitment and preparation of future chemistry teachers.

A focus on the early stages of education and on family attitudes and involvement is an important component of a comprehensive effort to improve U.S. science education.

The final session of the workshop focused on what participants thought should be done next. What actions should be taken, who should take those actions, and how could they be funded? Although opinions on some issues varied, participants at the workshop identified and elaborated on the need for several broad initiatives. These actions represent the opinions of a panel and of various workshop participants and do not represent consensus recommendations.

COORDINATING EFFORTS

"We have heard lots of examples of interesting projects that are going on and efforts that range from more formal professional development to more informal kinds of outreach projects," said Joan Prival of the National Science Founda-

tion (NSF). "What I keep thinking of is how could we get better coordination of all these wonderful projects? We can't have a conference like this every month."

Coordination is needed on many levels, Prival said. Activities have to be coordinated within an institution so that people know what others are doing. Sometimes an organization may be geographically distributed, as in the case of the schools in a district. For example, decisions about professional development or the structure of a school day can be made in one part of a district that significantly affect the daily lives of teachers throughout the district.

Coordination across institutions can encompass institutions that are similar or dissimilar. For example, coordination across industry, private foundations, colleges and universities, and schools is necessary to improve some aspects of chemistry education. "That's a big challenge," said Prival. One way to coordinate activities across organizations is through partnerships, which offer the prospect of "increasing the coherence and lowering the noise in the system," said Gerry Wheeler of the National Science Teachers Association.

Mentoring is a kind of partnership that can be especially valuable for teachers. "Teaching is a very lonely profession," said Bill Carroll of Occidental Chemical Corporation. "It's kind of like being a stand-up comedian and doing six shows a night. It's just you standing up there."

Also, chemistry teachers need to be recruited to participate in teacher outreach and professional development opportunities, Sandra Laursen of the University of Colorado at Boulder pointed out. Some teachers are willing to spend a Saturday or an entire week or two developing their skills for fairly little compensation, because they are interested and motivated. Yet many other teachers typically do not attend workshops or other professional development activities. Ways need to be found to attract this more representative group of teachers, especially to the more specialized activities that go beyond what can be done in large programs. One way to reach these teachers, suggested Prival, is to train teacher leaders who can reach out to all of the teachers in a school, both through site-based professional development and by encouraging teachers to become involved in activities outside the school. Wheeler suggested thinking more like a business in encouraging teachers and offering them incentives to attend professional development activities. "You have to give them a reason to show."

Out-of-school programs for students also need to be considered, said Rena Subotnik of the American Psychological Society, because these can have a profound effect on students. Many students who are excited about science take advantage of such programs, but many more students do not have access to them, especially in parts of the country where out-of-school programs are not located nearby.

IDENTIFYING AND DISSEMINATING SUCCESSFUL PRACTICES

Government agencies, including NSF and the National Institutes of Health (NIH), are strongly encouraging program designers and practitioners to include evaluation in the design and implementation of their programs. "You would never start on a research project without knowing what's going on in the field, without being aware of the literature, and without communicating with your colleagues," said Prival. Program leaders need to approach evaluation "seriously as a scholarly effort." In many cases, this will require partnerships between natural scientists, social scientists, and educators.

It is important to consider evaluation at the beginning of a program, because that influences the program's design, said Penny Gilmer of the National Association for Research in Science Teaching. "Then once you have started it, don't wait until the end to do summative evaluation. Do formative evaluation all during it to improve your program as it proceeds. This is really critical." In addition, evaluations should look not just at what worked but at what can be improved. "That can improve your next program and help others avoid similar pitfalls." A theoretical perspective is needed to inform the design of a program and its evaluation. For example, Gilmer recommended a recent book by George Bodner and MaryKay Orgill that is devoted to describing and critiquing the theoretical frameworks used by chemistry education and science education more broadly.[1]

A broader problem, according to Wheeler, is that education lacks a model of progress. Partly for this reason, data demonstrating the lack of effectiveness of one-day or two-day training sessions in science teaching do not keep such sessions from being offered. Education researchers need a way to determine what does not work. Wheeler said that principal investigators of programs that do not work should release information about the program and what went wrong so that others can learn from the experience. If such information cannot be released publicly, at least it can be discussed on listservs among individuals involved in designing and delivering programs. Education needs a "culture of criticism," said Wheeler.

In addition, inquiry-based teaching is a topic that needs greater investigation, Wheeler said. The idea has many different definitions and many ways of being implemented. It may not work in every situation with every child. There may be a role for other teaching approaches, including direct instruction. Furthermore, as Hai-Lung Dai of Temple University pointed out, inquiry-based instruction requires teachers who know the content well, which emphasizes the close connection between content knowledge and pedagogy. Prival emphasized that the outcomes of education are the ultimate objective, and there are different possible pathways to reach that objective.

Wheeler cautioned against setting up a dichotomy between inquiry and something else. "Most good teachers in fact have a whole bunch of tricks in their bags. When they are trying to break a misconception, that strategy is entirely different than when they are trying to do something abstract like control of variables with fourth graders. It's much more of a continuum."

Eric Jakobsson from the University of Illinois urged NSF and the Department of Education to invest funds in research to determine the best approaches to pedagogy for chemistry. Prival responded that the NSF is funding such research. She added that most people probably would not write up their failures, though some of that information may be in their annual reports.

---

[1]George M. Bodner and MaryKay Orgill. 2007. *Theoretical Frameworks for Research in Chemistry/Science Education*. Upper Saddle River, NJ: Prentice Hall.

The techniques of corporate America should be applied to the design, implementation, and evaluation of programs—including those sponsored by corporations, Wheeler said. If one plant is making progress and another is failing, businesses shut down the failing plant. "Nonprofit is a tax status," he said, "not a business plan." Educators need to find models that work and are sustainable, rather than seeking to do the same thing when results show that conventional approaches are not working.

Participants discussed strengthening teacher certification requirements but did not agree on what should be done. Some urged stricter requirements, whereas others said that such requirements can keep good teachers out of the classroom. Also, it is almost always possible for states to work around federal mandates.

Successful practices need not be identified only through formal and rigorous evaluations, said Prival. Things that people have learned in setting up partnerships and carrying out projects can be very valuable for someone engaged in a related activity. Ways need to found of sharing those practices. Papers, electronic journals, and other forms of publication are outlets for such information. For example, monographs with chapters written by participants in a program can provide valuable insights, but many other kinds of communication and forms of dissemination are possible. For example, a Web-based dissemination system could provide access to a wide variety of formal and informal resources. "If you have teachers in a Web site where they can interact with each other, they read each other's comments, they get ideas, and they have to put these ideas together, that's part of learning," said Gilmer. Prival mentioned that NSF could look for a grantee to provide such a service for chemistry education. Mary Kirchhoff of the American Chemical Society suggested that the ChemEd pathway of the National Science Digital Library could be the possible base of such a repository.

SUPPORTING TEACHERS AND TEACHING

Edward Crowe referred to several leverage points that can help produce change. First, more than 1,000 colleges and universities train teachers, and working with those institutions is essential to improve the effectiveness of the teaching force. Another approach is to identify bright and committed chemistry majors and work with them on their teaching skills. Teachers need solid preparation to be able to teach at high levels. Short-term workshops can get teachers excited and give them ideas to use in the classroom, but they cannot make up for a lack of content and pedagogical knowledge and training.

Second, if assessments were more sophisticated, teaching to the test would be a virtue. Crowe said, "What we could do is connect better learning outcomes and drive teaching toward those outcomes by having better assessment."

Finally, professional development sessions have the potential to waste billions of dollars. Most "have nothing to do with student learning and almost nothing to do with teaching quality," said Crowe. A much more important consideration may be hiring practices. Studies have demonstrated that major differences in teaching quality and student achievement can come from changing who is hired in a school district. Mentoring and induction programs also can make a big difference for new teachers. In addition, the leadership of individual schools is becoming more important, as many urban districts have begun to distribute authority, responsibility, and even budgets to individual principals.

Once teachers have participated in a professional development activity, they need continued support, said Prival. Teachers need help to implement what they have learned, which often requires follow-up work by an educational program. Also, isolation can occur in any school district. Teachers need enough experience to know that when they encounter a problem they can pick up a phone and find someone to help them solve it, said Hratch Semerjian of the Council for Chemical Research. Also, said Carroll, retirees around the country could help break down that isolation by assisting in schools, especially at the middle school or elementary school level.

Teachers need to know what content it is important to master both before entering teaching and while engaged in the profession and how best to convey content to students. However, content and pedagogy are often inseparable, as Kirchhoff pointed out. "The appropriate approach is just as important as the content, and they go hand in hand."

Disciplinary majors are proving to be rich pools of future teachers. If pedagogy courses can be integrated into the preparation of chemistry majors interested in teaching, either in four years or during a fifth year of preparation, they can become very effective teachers. This approach requires that college and university faculty acknowledge teaching as a valuable future profession for chemistry students, which "is a big change in the culture," said Prival. A further need is to professionalize chemistry teaching through better pay and other measures of respect, so that teachers have an identity both as chemists and as teachers. "It is very important for them to be able to be treated as professionals and have time in their work space for them to communicate, to visit each other's classrooms, and to share what they are doing."

An important way to support teachers is to provide them with pre-service education that can be applied directly in the classroom. Pre-service teachers also need the direct involvement with research that has proven valuable for in-service teachers, said Prival. If teachers come into the profession with strong content knowledge, they will need less professional development than would be the case otherwise.

BUILDING A BASE FOR SUCCESS

One approach to solving a problem is to build backward to its source, said Dai. Today, more than half of the chemistry graduate students in U.S. colleges and universities are from abroad. Also, knowledge and understanding of chemistry are low among the general public. Why do other cultures do a much better job of science education while spending less money on education? Dai felt that a cross-cultural study might be able to determine the answer.

Dai suggested two solutions to the problem of inadequate chemistry education in the United States: going back to basics and starting earlier. If the problem is in high school, the solution may lie in middle school or elementary school. Faculty at Temple have found that more students arrive having taken advanced placement (AP) classes and more arrive needing remedial education. In many cases, this remedial education is essential even for students in science classes who need help with basic mathematics and reading to succeed. Temple has recently started administering mathematics placement tests to all incoming students, and about 15 percent of the incoming students fail at a ninth-grade level of mathematics. "This is the state of American education," said Dai.

Finally, Marshall Lih from NSF said that more direct involvement and influence on families will be necessary to solve some of the educational issues America faces. Parents and families have such a strong influence on the knowledge and attitudes of their children that the issue cannot be ignored. Talking about families is "a very sensitive issue," Lih acknowledged, but "family is important, and parental guidance is very important."

Appendixes

A

Workshop Agenda

MONDAY, AUGUST 4, 2008

8:00 a.m.-8:15 a.m.   *Welcome & Introductions*
Workshop Organizers:
Mark J. Cardillo, Camille & Henry Dreyfus Foundation
William F. Carroll, Occidental Chemical Corporation
Alex Harris, Brookhaven National Laboratory

Session 1: What Are the Major and General Issues in High School Chemistry Education?

State of Science—Chemical Education

8:15 a.m.   *STEM Education Overview*
What are the national indicators of student performance and teacher quality?
Kathryn D. Sullivan, Vice Chair of the National Science Board and Director, Battelle Center for Mathematics & Science Education Policy.

9:15 a.m.   *Importance of Teachers*
What influences high school student performance and interest in pursuing STEM (and chemistry in particular) degrees?
Robert H. Tai, University of Virginia

On the Frontlines—Teaching High School Chemistry

10:30 a.m.   *National Perspective*
Gerald Wheeler, National Science Teachers Association

11:00 a.m.   *State Perspective*
Roxie Allen, Associated Chemistry Teachers of Texas (ACT2)

11:30 a.m.   *Teacher Panel—Local Perspectives*
                              Caryn Galatis, Thomas A. Edison High School, Virginia
                              Brian J. Kennedy, Thomas Jefferson High School for S&T, Virginia
                              Kiara Delle Hargrove, Baltimore Polytechnic Institute, Baltimore, Maryland

Session 2: Who Is Doing What with Respect to High School Chemistry Education (and How Is Effectiveness Measured?)

Publicly Funded Programs

1:15 p.m.   National Institutes of Health
                              L. Anthony Beck, Division for Clinical Research Resources, National Center for Research
                              Resources

1:45: p.m.   National Science Foundation
                              Katherine Covert, Chemistry Division
                              Joan Prival, Division of Undergraduate Education

2:15 p.m.   Department of Energy-Academies Creating Teacher Scientists
                              Jeffery Dilks, Office of Workforce Development for Teachers and Scientists

2:45 p.m.   National Laboratory Perspective
                              Kenneth White, Office of Educational Opportunity, Brookhaven National Laboratory

Chair: Mark Cardillo, Camille & Henry Dreyfus Foundation

3:30 p.m.   ChemEd Conferences
                              Irwin Talesnick, Queens University, Canada

4:00 p.m.   University of Pennsylvania Science Teacher Institute
                              Constance Blasie and **Michael Klein,** University of Pennsylvania

4:30 p.m.   AirUCI Summer Workshop for Teachers
                              Sergey Nizkorodov, University of California-Irvine

5:00 p.m.   "Terrific Science" 25 Years of Outreach in Chemical Education
                              A summary of what works, what doesn't, and how we know.
                              Gil Pacey, Miami University

Poster Session: Sampling of Teacher Outreach Programs (invited and contributed)
5:30-7:30 p.m.

TUESDAY, AUGUST 5, 2008

8:00 a.m.   Welcome

Session 3: Who Is Doing What with Respect to High School Chemistry Education (and How Is Effectiveness Measured?)

Privately Funded Programs

8:15 a.m.   A Corporate Perspective
Bridget McCourt, Bayer Corporation, Making Science Make Sense

8:45 a.m.   A Foundation Perspective
Bryce Hach, Hach Scientific Foundation

9:15 a.m.   What Are Foundations Doing?
Sandra Laursen, University of Colorado

9:45 a.m.   What Are Nonprofit Organizations Doing?
Reeny D. Davison ASSET (Achieving Student Success Through Excellence in Teaching) Inc./ Science: It's Elementary

10:30 a.m.   **Patricia M. Soochan**, Howard Hughes Medical Institute

11:00 a.m.   Workshop Wrap-up with the "Action" Panel
William F. Carroll, *Moderator*
- **Joan Prival**, NSF Division of Undergraduate Education
- **Mary M. Kirchhoff**, American Chemical Society
- **Penny J. Gilmer**, National Association for Research in Science Teaching
- **Gerald Wheeler**, National Association of Science Teachers
- **Hai-Lung Dai**, Temple University

12:00 p.m.   Workshop Adjourns

B

Biographies

ORGANIZERS

Mark J. Cardillo is the executive director of the Camille & Henry Dreyfus Foundation. Dr. Cardillo received his bachelor of science degree from Stevens Institute of Technology in 1964 and his Ph.D. degree in chemistry from Cornell University in 1970. He was a research associate at Brown University, a CNR research scientist at the University of Genoa, and a PRF research fellow in the Mechanical Engineering Department at the Massachusetts Institute of Technology. In 1975, Dr. Cardillo joined Bell Laboratories as a member of the technical staff in the Surface Physics Department. He was appointed head of the Chemical Physics Research Department in 1981 and subsequently named head of the Photonics Materials Research Department. Most recently, he held the position of director of Broad Band Access Research. Dr. Cardillo is a fellow of the American Physical Society. He has been the Phillips lecturer at Haverford College and a Langmuir lecturer of the American Chemical Society. He received the Medard Welch Award of the American Vacuum Society in 1987, the Innovations in Real Materials Award in 1998, and the Pel Associates Award in Applied Polymer Chemistry in 2000.

William F. Carroll is vice president of Occidental Chemical Corporation in Dallas, Texas, and an adjunct industrial professor of chemistry at Indiana University. He served as American Chemical Society (ACS) president in 2005 and as a member of the ACS Board of Directors from 2004 to 2006. He is the former chair of International Activities Committee at ACS. He earned a B.A. from DePauw, an M.S. from Tulane University (1975), and a Ph.D. from Indiana University (1978). Carroll has been an ACS member since 1974 and has served on a number of committees. He holds memberships in the Society of Plastics Engineers; American Association for the Advancement of Science; National Organization for the Professional Advancement of Black Chemists and Chemical Engineers; and National Fire Protection Association; he was the recipient of the Vinyl Institute Roy T. Gottesman Leadership Award in 2000.

Alex Harris is chair of Brookhaven National Laboratory's Chemistry Department. Dr. Harris earned a B.A. in chemistry from Swarthmore College in 1978 and a Ph.D. in physical chemistry from the University of California at Berkeley in 1985. He joined AT&T Bell Laboratories, Murray Hill, New Jersey (now Bell Laboratories, Lucent Technologies), in 1985 as a member of the technical staff, Chemical Physics Research, and became head of the Materials Chemistry Research Department in 1996. In 2000, he joined Agere Systems, Allentown, Pennsylvania, as director of the Guided Wave and Electro-optics Research Department, a position he held until he came to Brookhaven in 2003.

SPEAKERS

Roxie Allen teaches science in the Upper School of St. John's School in Houston, Texas. She was first appointed as a teacher at the school in 1990 after obtaining a B.S. from Texas A&M University and an M.S. from the University of Houston. She is also the past president of the Association of Chemistry Teachers of Texas (ACT2). Roxie's dedication to chemistry education extends beyond her classroom; she recently completed her three-year term as a mentor with the U.S. Chemistry Olympiad. From 2004 to 2006, Roxie helped direct the lecture and laboratory components of the U.S. Chemistry Olympiad Camp attended by the 20 top scorers on the national exam.

L. Anthony Beck received his Ph.D. in molecular biology and biochemistry from the University of California, Irvine, and Brookhaven National Laboratory and his postdoctoral training in Denver at both the University of Colorado Health Sciences Center on the molecular biology of brain development and the Eleanor Roosevelt Institute for Cancer Research on the posttranslational processing and nuclear targeting of hepatic and viral proteins. In 1990, he was hired by Life Technologies, Inc. (LTI), in Gaithersburg, Maryland, to establish its Molecular Biology and Cell Culture Training Center. In 1992, he moved to Cellco, Inc., a hollow-fiber bioreactor company based in Germantown, Maryland, where he held managerial positions in Research Applications, Drug Discovery, and Asia Pacific Business Development. In 1997 and 1998, Dr. Beck was a consultant for Walter Reed Army Medical Center and the American Registry of Pathology on protocol development for hollow fiber-based zero-gravity cell culture experiments for the National Aeronautics and Space Administration's (NASA's) Space Shuttle program. In 1998, he co-founded Tissue Engineering Sciences (TES), Inc., where he served as vice president for Research & Development (R&D). TES' R&D portfolio included bioartificial blood vessels, ex vivo arterial perfusion models, and in vitro blood-brain barrier and pharmacokinetic systems. In 2000, Dr. Beck joined the National Institute on Alcohol Abuse and Alcoholism (NIAAA) as a scientific review administrator; he moved to the National Center for Research Resources (NCRR) at the National Institutes of Health (NIH) in 2002 where his programmatic responsibilities include the trans-NIH R24 Human Embryonic Stem Cell Infrastructure Awards, the S07 Human Subjects Research Enhancement Program, M01 General Clinical Research Centers, and the R25 Science Education Partnership Award.

Constance Blasie is the program director of the University of Pennsylvania's Science Teacher Institute (STI), which offers a master of chemistry education program and a master of integrated science education program to improve the science content knowledge of in-service science teachers. Since her retirement in 1995 from a 30-year career as a secondary-level mathematics teacher, department chair, and curriculum developer in suburban Philadelphia, Blasie has been instrumental in the design, development, and implementation of the two STI master's degree programs. She is a graduate of the University of Michigan.

Katharine Covert is the program director for Chemistry Centers and Special Projects at the National Science Foundation (NSF) Division of Chemistry. She joined the division in 2001 and has worked in many programs, including the Inorganic Program; Collaboratives, Environmental Molecular Science Institutes; Discovery Corps Fellows; Research Experience for Undergraduates; and now the Chemistry Centers. Kathy did her undergraduate work at the College of William and Mary (B.S., 1985) and her graduate work at Cornell University (Ph.D., 1991) and then went to the University of Oregon for a postdoctoral position. She taught at West Virginia University and Bates College before moving to NSF.

Hai-Lung Dai became the dean of the College of Science and Technology at Temple University in January 2007. Previously, he was the Hirschmann-Makineni Professor, chair of Chemistry, and founding director of the Science Teacher Institute at the University of Pennsylvania. Dai came to the United States for graduate study in chemistry in 1976 at the University of California, Berkeley, after graduation from the National Taiwan University and military service. After a postdoctoral stint at the Massachusetts Institute of Technology (MIT) he arrived at Penn as an assistant professor in 1984. Dai was promoted to full professor in 1992 and was the chairman of the Chemistry Department from 1996 to 2002, during which time he established the M.S. in chemistry education program that has trained more than 100 in-service high school chemistry teachers. Under his leadership at Temple University's College of Science and Technology, in collaboration with the College of Education, he established the TUteach program aimed at attracting math and science majors for pedagogical training to become content-prepared math and science teachers. As an accomplished researcher, he has published more than 150 papers in the areas of molecular and surface sciences and received numerous honors including a Dreyfus Foundation Teacher-Scholar Award, a Sloan Fellowship, the Coblentz Prize in Molecular Spectroscopy, the Morino Lectureship (Japan), a Humboldt Fellowship (Germany), the American Chemical Society Philadelphia Section Award, and a Guggenheim Fellowship and the Ellis Lippincott Award in Spectroscopy. He is a fellow of the American Physical Society and was elected by the membership to be the chair of the Chemical Physics Division of the American Physical Society in 2006.

Reeny D. Davison is the executive director of ASSET (Achieving Student Success Through Excellence in Teaching) Inc., a nonprofit organization that works to continuously improve teaching and learning through science education. She earned a B.A. in German and English from San Jose State University and spent the junior year studying abroad at the Free University in Berlin. She earned an M.A. in German literature and cultural history and a TESOL (Teaching English to Speakers of Other Languages) certificate from the University of Pittsburgh in 1989. Reeny also received an Ed.D. in educational leadership from Duquesne University in 2000. After working at McKinsey and Company, Inc., for two years she began her teaching career in the Netherlands and taught at the college and adult level for more than 20 years. She employs both her education and her business skills to ensure ASSET's entrepreneurial growth. She has received several awards for her work at ASSET from institutions such as Duquesne University and Carlow College.

Jeffery Dilks is a staff member in the Office of Workforce Development in the Office of Science of the Department of Energy. He also serves as editor of the Department of Energy's *Journal of Undergraduate Research*. He has a B.A. in physics from the University of Illinois and an M.S. in the history of science and technology from Illinois State University (ISU). During his time as a physics teacher at Ames High School in Iowa, he was one of 24 science teachers chosen for the Quark-Net project in 1999. The project aimed to expose high school teachers to the experiments being conducted and was successful; Dilks built a new Cerenkov calorimeter for use at the European Organization for Nuclear Research (CERN) Large Hadron Collider. Dilks was named a 2006-2007 Albert Einstein fellow.

Caryn Galatis teaches chemistry at Thomas Edison High School in Fairfax, Virginia. Galatis earned a B.S. in chemistry from Mary Washington College and an M.Ed. from the University of Virginia. She has been a science and math teacher in Fairfax County Public Schools for 30 years, teaching primarily chemistry. Galatis has taught all levels of chemistry, general through advanced placement and international baccalaureate, and has been the Science Department chair at Edison. since 1989. Besides her teaching responsibilities, Galatis has been very involved in curriculum and staff development work both in Fairfax County and in other parts of Virginia. In the summers she teaches an online chemistry course and works on Standards of Learning content review for the State of Virginia. In 1991, Galatis was selected as chemistry teacher of the year by the American Chemical Society.

Penny J. Gilmer is a professor in the Department of Chemistry and Biochemistry at Florida State University (FSU). Gilmer received her Ph.D. in biochemistry from the University of California, Berkeley, and held two fellowships at Stanford until joining the FSU faculty in 1977. In her quest to be a "lifelong learner," Professor Gilmer earned her D. Sc.Ed. in science and mathematics education from Curtin University of Technology in 2004. Currently, her primary research interests lie in science education. Professor Gilmer has been recognized for her "innovative research and teaching on how to bring science and technology, particularly ethics in science, to students and the community" by the American Association for the Advancement of Science (AAAS). Professor Gilmer is a mentor to both students and teachers, encouraging the use of action research to evaluate areas for improvement in teaching and learning. She also serves as the principal investigator of an FSU subcontract for a project entitled "Science Collaboration: Immersion, Inquiry, Innovation," with the Panhandle Area Educational Consortium, funded through the State of Florida. She is the co-editor of *Transforming Undergraduate Science Teaching: Social Constructivist Perspectives* (Peter Lang Publishing, Inc., 2002).

Bryce Hach is the executive director of the Hach Scientific Foundation and a former high school science teacher. Since 2005, the Hach Foundation has focused on chemistry education from kindergarten to high school. To strengthen the field of science education, the Second Career Chemistry Teacher Scholarship was established in 2007 to encourage career chemists to become chemistry teachers. Hach holds a bachelor's degree in history and biology and a master's in public policy management.

Kiara Delle Hargrove strives to motivate urban high school students in chemistry as a science teacher at Baltimore Polytechnic Institute, one of Maryland's top-performing high schools. Hargrove frequently turns lessons into fun, active experiments, such as her demonstration about distilling water from a can of soda, which became a competition to see who could distill the most water. She integrates reading and writing strategies into her lessons, insisting that the composition of her students' science papers be as accurate as the science and math. Teaching a variety of academic levels simultaneously, from special education to gifted-level courses, she differentiates instruction to reach every student. Hargrove facilitates remedial math and science study skills among incoming freshmen through the Summer Bridge Program, and serves as the ninth-grade adviser. As co-adviser for the Math Engineering and Science Association (MESA), she helps elevate the study of math and science among girls, especially African Americans, at Sudbrook Magnet Middle School. Hargrove was chair of the School Improvement Team from 2006 to 2007 and is coauthor of the School Improvement Plan. She has influenced many of her fellow teachers to go beyond traditional approaches to teaching. In 2007, she was one of the 75 recipients of the Milken National Educator Award.

Brian J. Kennedy teaches chemistry and is the director of the Chemical Analysis Research Laboratory at Thomas Jefferson High School for Science and Technology (TJHSST). He holds a Ph.D. in analytical chemistry from the University of Wyoming (1997) and a B.S. in chemistry and B.S. in physical science from Radford University. He is currently enrolled in a graduate education leadership program at George Mason University. Prior to teaching at TJHSST, Kennedy taught science for three years through Teach for America and also completed several years as a National Research Council (NRC) postdoctoral research assistant at the U.S. Army Research Laboratory, Aberdeen Proving Grounds, in Maryland. During the last seven years at TJHSST, Kennedy has taught all levels of chemistry and sponsors the school's

Chemistry Olympiad Team. Kennedy is the recipient of the American Chemical Society Capitol Society of Washington 2008 Leo Schubert Memorial Award for the Outstanding Teaching of High School Chemistry.

Mary M. Kirchhoff is the director of the American Chemical Society Education Division. She holds a Ph.D. in organic chemistry from the University of New Hampshire, an M.S. degree in chemistry from Duquesne University, Pittsburgh, Pennsylvania, and a B.A. in chemistry from Russell Sage College, Troy, N.Y. Kirchhoff served as assistant director for special projects in the Education Division and was assistant director of the ACS Green Chemistry Institute for three years, where she managed day-to-day operations of the institute. Prior to joining ACS, she worked at the U.S. Environmental Protection Agency and was an associate professor and an assistant professor of chemistry at Trinity College in Washington, DC. In 2007, Kirchhoff was named a AAAS fellow, "for leadership in promoting the environmentally sound practice of green chemistry in education and research." Kirchhoff is a coauthor of *Designing Safer Polymers* (Wiley-IEEE, 2000) and co-editor of *Greener Approaches to Undergraduate Chemistry Experiments* (ACS, 2002).

Michael Klein, professor of chemistry and physical sciences and director of Penn Laboratory for Research on the Structure of Matter, was cited by the National Academy of Sciences for work that has led to physically significant and predictive descriptions of hydrogen-bonded liquids, self-assembled monolayers, supercooled liquids, conducting fluids, and biological membranes. Klein has devised computational methods to predict how the properties of matter respond to changes in pressure and temperature and is noted for his computer simulations of molecular materials. Klein, who has authored approximately 500 papers in research journals, ranks as the world's 96th most-cited chemist, according to an Institute for Scientific Information analysis of research papers published from 1981 to 1997. He has edited three books and serves on the editorial boards of numerous journals. He was a Guggenheim fellow in 1989-1990 and is a fellow of the Royal Society of Canada, the Chemical Institute of Canada, and the American Physical Society. Klein joined the Penn faculty in 1987 after 19 years at the National Research Council Canada (NRCC), culminating as principal research officer in the NRCC Chemistry Division. He received a B.Sc. in 1961 and Ph.D. in 1964 from the University of Bristol in the United Kingdom.

Sandra Laursen is co-director of Ethnography & Evaluation Research (E&ER), an independent research unit at the University of Colorado at Boulder. E&ER is an interdisciplinary team that conducts research and evaluation studies of education and career paths in science, engineering, and mathematics. Recent projects have examined the advance-

ment of academic women scientists, programs to enhance the success of minority science students, outreach programs in biology and geology, and a multicampus initiative to improve undergraduate mathematics education. A new study is investigating graduate education and career preparation in chemistry, and a forthcoming book discusses the outcomes of undergraduate research apprenticeships in the sciences. In addition to her research and evaluation work, Laursen is an outreach scientist at the Cooperative Institute for Research in Environmental Sciences, where she leads courses and workshops on earth science and physical science and inquiry-based teaching methods for K-12 teachers, college instructors, and scientists involved in outreach. She has a Ph.D. in physical chemistry, with research experience in photochemistry, free radical reactions, and atmospheric chemical kinetics.

Bridget McCourt is the Director of Bayer Corporation's Making Science Make Sense® science literacy initiative. Prior to joining Bayer in 2006, she worked as communication representative at NOVA Chemicals. Bridget earned her B.A. in history from St. Mary's of Notre Dame in 1993.

Sergey Nizkorodov is an associate professor of chemistry at the University of California, Irvine (UCI). He earned his M.S. degree in biochemistry at Novosibirsk State University and his Ph.D. degree in physical chemistry at the University of Basel. Professor Nizkorodov is the principal investigator in the Aerosol Photochemistry Group (*http://aerosol.chem.uci.edu*), a component of the NSF-funded AirUCI institute. His research focuses on the interaction between solar radiation and atmospheric aerosols and on indoor air pollution. In 2005, he was awarded the Coblentz Award as an outstanding young molecular spectroscopist. He is a recipient of the 2007 Camille Dreyfus Teacher-Scholar Award and the 2006 UCI School of Physical Sciences Award for Outstanding Contributions to Undergraduate Education for his educational work at UCI.

Gil Pacey is currently leading the Miami University Nanotechnology Initiative that is charged with incorporating nanotechnology into the teaching and research of Miami University. His current research efforts focus on nanotechnology and microfluidics in order to develop a "smart nozzle" technology, in which the nozzle is capable of detecting the components of the substance being pumped through and providing necessary feedback to the controlling system. Pacey has served on the faculty of Miami University of Ohio since 1979, and currently serves as both the associate dean for Research and Scholarship and the director of the Miami University Center for Nanotechnology within the Department of Chemistry and Biochemistry. He previously served as the director of the Ohio Micromachining Analytical Chemistry Consortium (1997-2001). Professor Pacey received his Ph.D.

in 1979 from Loyola University of Chicago, where his graduate and postgraduate adviser was Carl E. Moore. The author of more than 100 publications, Professor Pacey has eight years of experience as an industry consultant.

Joan Prival is the lead program director for the Robert Noyce Teacher Scholarship program in the Division of Undergraduate Education at the National Science Foundation. In addition, she serves as a program director in the Math and Science Partnership program and the Advanced Technological Education program. She received a B.A. degree in biological sciences from Wellesley College and a Ph.D. in biochemistry from the Massachusetts Institute of Technology. As a research biochemist, she conducted studies on blood cell differentiation and leukemia at the National Cancer Institute. Prior to coming to NSF in 1997, she served as an education policy specialist for 14 years with the Washington DC, public schools. In 1999 she was awarded a fellowship from the Japan Society for Promoting Science to study teacher preparation in Japan. She has received four NSF Director's Awards, including the NSF Director's Award for Superior Accomplishment in 2002.

Patricia M. Soochan received a bachelor and a master of science degree from George Washington University in 1977 and 1981, respectively. In 1982 she became a biochemist at Bethesda Research Labs, later to be known as Life Technologies. Her work included conducting biotechnology workshops in France and Brazil. In 1987 she became a senior information specialist at Social and Scientific Systems, a consultant to the National Cancer Institute. There, she worked with physicians in preparing reports of investigational cancer therapies. In 1991 she joined the National Science Foundation as a science assistant-biologist involved in grants management in the cell biology program. In 1994 she joined the undergraduate science education program at the Howard Hughes Medical Institute, where she is now a program officer engaged in all aspects of competition and award management from system design to policy development, with an emphasis on college grantees.

Kathryn D. Sullivan was named director, Battelle Center for Mathematics and Science Education Policy at the John Glenn School of Public Affairs, Ohio State University, Columbus, in October 2006. The center addresses the nation's global competitiveness by developing policies and practices to increase the number of students in the science, technology, engineering, and mathematics fields. Sullivan previously served as president and chief executive officer of the Center of Science and Industry (COSI), a dynamic center of hands-on science learning, where she now volunteers as a science adviser. Prior to joining COSI, Sullivan was the chief scientist of the National Oceanic and Atmospheric Administration (NOAA). Sullivan is a veteran of three Space Shuttle mis-

sions and the first American woman to walk in space. She holds a bachelor of science degree in earth sciences from University of California at Santa Cruz and a Ph.D. in oceanography from Dalhousie University (Nova Scotia). She was appointed to the National Science Board in 2004 and elected vice chairman in 2006.

Robert H. Tai is an associate professor in the Curry School of Education at the University of Virginia. After receiving a B.A. and B.S. in mathematics and physics (1986) from the University of Florida, Professor Tai went on to earn his M.S. in physics from the University of Illinois in 1987. After working as a research assistant in the Nuclear Physics Laboratory at the University of Illinois, Professor Tai taught physics in Illinois and Texas. Professor Tai earned his Ed.M. (1994) and Ed.D. (1998) in science education from the Harvard University Graduate School of Education, where he then worked as a researcher and teaching fellow. Professor Tai has taught 15 college courses on science education between his previous position at the College of Staten Island and his current position at the University of Virginia. In May 2008, Professor Tai was recognized with the 2008 Award for Education Research Leadership from the Council of Scientific Society Presidents for his widely cited research into the factors that lead students to become scientists.

Irwin Talesnick is a professor emeritus at Queen's University in Ontario. He continues to create and distribute educational and fascinating demonstrations through his company, S17 Science Supplies and Services. His experiences include a lifetime of teaching, of training teachers, of providing educational materials for others to use, and of giving workshops. Starting in 1960, he taught high school chemistry, physics, and general science in Toronto. Then for 25 years he was a professor of chemical education at the Faculty of Education at Queen's University in Kingston, Ontario, preparing new teachers for a life in the classroom. Talesnick has been the recipient of the Science Association of Ontario's (STAO/APSO) Life Member and Service Awards. In 1993, the year he retired from Queen's University, the STAO/APSO Excellence in Teaching Award was replaced with the Irwin Talesnick Award for Excellence in the Teaching of Science. Since his retirement, he has expanded his workshop and lecturing schedule, which over the years has taken him from Canada to the United States, Mexico, England, Wales, China, Sweden, and Israel. Irwin was the chair of the ChemEd conferences in 1987 and 1989 at Queen's University in Kingston, and in 2001 at York University in Toronto. He has been a presenter at all of the ChemEd conferences since their beginning in 1973 at Waterloo.

Gerald Wheeler joined as executive director of the National Science Teachers Association in 1995. He received an undergraduate degree in science education from Boston University and a master's degree in physics and a Ph.D. in experimental nuclear physics, both from the State University of New York at Stony Brook. Between undergraduate and graduate school, he taught high school physics, chemistry, and physical science. For much of his career, Dr. Wheeler has played a key role in the development of mass media projects that showcase science for students. Prior to joining the National Science Teachers Association, Dr. Wheeler was director of the Science/Math Resource Center and professor of physics at Montana State University. He also headed the AAAS Public Understanding of Science and Technology Division and has served as president of the American Association of Physics Teachers. He is a fellow of the W. K. Kellogg Foundation and AAAS and has served on advisory boards and committees for the American Institute of Physics and the National Assessment of Educational Progress.

Kenneth White has served as manager of the Office of Educational Programs (OEP) at the U.S. Department of Energy's (DOE's) Brookhaven National Laboratory since 2004. White earned a B.S. with concentrations in engineering technology and education from the University of the State of New York, Regents College, Albany, in 1990 and an M.B.A. from Dowling College in 2003. From 1978 to 1986, he served in the U.S. Navy as a nuclear training instructor, a lead engineering laboratory technician, and an engineering watch supervisor. In 1987, he became supervisor of Training Program Development for the Long Island Lighting Company, and in 1990, he joined Brookhaven Lab as a senior reactor support specialist at the High Flux Beam Reactor (HFBR). In 1994, he became leader of the Water Chemistry Group at the HFBR. In 1998, White was appointed as the special assistant to the assistant laboratory director for Community, Education, Government and Public Affairs (CEGPA) and manager of Environmental Management Community Relations within CEGPA. In addition to serving as interim OEP manager since December 2003 until his appointment as manager, he also filled that position from 2000 to 2001. A past president of the Long Island Section of the American Nuclear Society, White is the recipient of the American Nuclear Society Training Excellence Award and the Brookhaven Award for distinguished service to the laboratory.

C

Poster Abstracts

The Penn Master of Chemistry Education Program: Data from Cohorts 1-5

Jane Butler Kahle,[1] Yue Li,[2] Constance Blasie[3]

[1]*Ohio's Evaluation & Assessment Center for Mathematics and Science Education, Miami University, McGuffey Hall, Oxford, OH 45056; e-mail: kahlejb@muohio.edu*

[2]*Ohio's Evaluation & Assessment Center for Mathematics and Science Education, Miami University, McGuffey Hall, Oxford, OH 45056; e-mail: liy@muohio.edu*

[3]*Penn Science Teacher Institute, University of Pennsylvania, 231 S. 34th Street, Philadelphia, PA 19104-6323; e-mail; cwblasie@sas.upenn.edu*

The University of Pennsylvania's Master of Chemistry Education (MCE) project graduated five cohorts of approximately 20 teachers between 2002 and 2006. One year after teachers in the last cohort earned their degrees, the Penn Science Teacher Institute (Penn STI) initiated a follow-up study to ascertain if the goals of the MCE project had been sustained. For example, were the teachers incorporating updated content knowledge into their lessons and were their students learning more chemistry? A total of 74 of the 82 graduates participated in some aspect of this study. Because baseline data were not available for the MCE teachers and their students, baseline data from a comparable group of chemistry teachers enrolled in the first cohort of the Penn STI project and their students were used in some analyses. Among other findings, the data indicate that MCE met its goals: (1) to reach urban teachers and teachers with limited chemistry knowledge; (2) to increase the use of inquiry-based instruction; and (3) to improve student achievement in chemistry (students of MCE graduates scored significantly higher than the comparison group).

Mechanical Resonance Characteristics in a Borate Polymer Environment as a Function of Glucose Concentration—A Student-Friendly Application of Chemical Engineering in the High School Science Classroom

Ellen M. Johnson,[1, 2] Loraine P. Snead,[1, 2] Annette D. Shine[2]

[1]*Wilmington Friends School, Wilmington, Delaware*

[2]*Department of Chemical Engineering, University of Delaware, Newark*

As a preliminary model for in vivo detection of glucose levels in diabetic patients, using remote sensing, we have developed a bench-top system for analysis of the relationship between the glucose concentration in a polymer containing borate and hydroxyl groups. We have used the audio editing program Amadeus Pro (HairerSoft.com) for analysis of properties of waveforms created by standard tuning forks suspended into a polymer-glucose solution. In addition to direct application to the chemical principles regarding the replacement of the polymer hydroxyl groups by the hydroxyl groups in glucose molecules, this general method can be extended in the interdisciplinary, inquiry-based classroom. Students can design further experiments testing multiple input variables and consider the contributions to various science disciplines and applications of related mathematical principles as seen in the data analysis. This work is an outgrowth of our association with the University of Delaware-Nature InSpired Engineering Research-Experience for Teachers (UD-NISE-RET) program in the summer of 2008 (*http://www.nise.udel.edu*).

PPG R&D Science Education Council: ENGAGE—EMPOWER—ENRICH

Kimberly Schaaf

PPG Industries Coatings Innovation Center, 4325 Rosanna Drive, Allison Park, PA 15101; e-mail: kschaaf@ppg.com

The mission of the PPG Science Education Council is to encourage and facilitate the participation of PPG associates in programs that educate our communities in sciences and engineering and inspire students to pursue scientific professions. We recognize that in order to fulfill the latter part of our mission we must also reach out to the educators that will be teaching those students. This poster highlights several teacher outreach programs currently in place as well as some of the other interactive and exciting activities sponsored by our group.

Summer Research Fellowships for Teachers: A Proven Model of Professional Development

Kaye Storm

Stanford University, Building 60, Room 214, Stanford, CA 94305-2063; e-mail: kstorm@stanford.edu

The Office of Science Outreach (OSO) at Stanford University has a long history of partnering with a San Francisco Bay area educational nonprofit to provide chemistry teachers unique professional development during the summer. Since 2005, 55 high school science teachers (including 22 teachers of chemistry) have held eight-week research fellowships within the university's science, engineering, and medical school labs. The teachers have cumulatively reached an estimated 21,000 students, more than one-third of whom are from groups that are underrepresented in the chemical sciences. The poster presents evidence that these teacher research fellowships result in greater teacher retention, motivation, and competency and that student standardized test scores and participation in extracurricular science activities increase following the teachers' experience.

The North Carolina School of Science and Mathematics Chemistry Faculties Outreach Efforts

Myra J. Halpin

North Carolina School of Science & Mathematics, 1219 Broad Street, Durham, NC 27705; e-mail: halpin@ncssm.edu

The North Carolina School of Science and Mathematics is a state-funded residential high school for students with high aptitudes in math and science. Part of our legislative mandate is to help improve the math and science education in the state. This poster describes the Chemistry Department's efforts to help North Carolina students and teachers statewide by providing (1) teacher workshops via our two-way audio and video distance learning program and honors and advanced placement (AP) chemistry online to small schools that do not have sufficient enrollment to offer advanced courses; (2) leadership in chemistry curriculum development, state-wide objectives, and end-of-course questions by working with NC-DPI; (3) summer residential workshops for North Carolina chemistry teachers to improve teachers' content knowledge and provide numerous laboratory activities that are easy and inexpensive for teachers to add to their existing program; (4) summer research opportunities for students to conduct research projects, RECAP, and RSI; (5) sessions at American Chemical Society (ACS), ChemEd, NCSTA, and NSTA meetings; and (6) animations and videos for teacher use via the Web site *www.dlt.ncssm.edu/TIGER*.

Pharmacology Education Partnership II: Teaching Neuroscience and Pharmacology to High School Students Improves Achievement in Biology and Chemistry*

Rochelle D. Schwartz-Bloom,[1] Myra J. Halpin,[2+] Jerome P. Reiter[3]

[1]Department of Pharmacology & Cancer Biology, Duke University Medical Center; e-mail: schwa001@duke.edu

[2]North Carolina School of Science & Mathematics; e-mail: halpin@ncssm.edu

[3]Department of Statistics & Decision Sciences, Duke University, Durham, N.C.; e-mail: jerry@stat.duke.edu

*Supported by a NIDA Science Education Drug Abuse Partnership Award # DA 10904.

[+]Presenting author.

The Pharmacology Education Partnership (PEP) is a curriculum developed for high school teachers, providing them with tools to teach the principles of biology and chemistry in the context of pharmacology (e.g., drugs of abuse) and the brain. We hypothesized that high school students might learn basic concepts in biology and chemistry better if the material is presented in the context of something interesting and relevant to their own lives. The PEP project includes several components such as curriculum design (six pharmacology modules), science content, professional development, and student assessment. In our first study, 50 teachers across the United States participated in a five-day workshop and field-tested the PEP curriculum in their classrooms; 4,000 of their students were tested and showed improvement in biology and chemistry compared to the standard curriculum (Schwartz-Bloom and Halpin, 2003). In this expanded study, 237 teachers were provided six hours of professional development in pharmacology and neuroscience at an NSTA meeting or via Distance Learning (two-way audio-video broadcasts). More than 10,000 students were tested for knowledge of basic biology and chemistry principles as well as advanced knowledge about drugs. The use of the PEP modules demonstrated significant gains in high school biology and chemistry classrooms using the PEP modules compared to the standard curriculum (Kwiek et al., 2007). The PEP curriculum can be accessed online at *www.thepepproject.net.*

References

N. C. Kwiek, M. J. Halpin, J. P. Reiter, L. A. Hoeffler, and R. D. Schwartz-Bloom. 2007. *Science* 317:1871-1872.

R. D. Schwartz-Bloom and M. J. Halpin. 2003. *J. Res. Sci. Teach.* 40:922-938.

Chemistry Institutes: Enhancing Science Teachers' Capacity and Curricula Using Trained Student Support

Michael F. Z. Page,[1] Edward D. Walton,[1] Joelle Opotowsky,[1] Laurie Riggs,[1] Brenda L. Oldroyd[2]
[1]*California State Polytechnic University, Pomona, CA 91768; e-mail: mfpage@csupomona.edu*
[2]*Diamond Ranch High School, Pomona, Calif.*

High school chemistry teachers are faced with tremendous challenges in teaching science to our students. According to the National Academies *America's Lab Report*, "Improving high school science teachers' capacity to lead laboratory experiences effectively is critical to advancing scientific educational goals." At Cal Poly Pomona, we have developed an innovative teacher-student program that couples high school science teachers with trained student teaching assistants, thereby increasing the teaching capacity of the instructor and allowing the class to perform more laboratory experiments. During our summer institutes the teachers and students work as a team to develop inquiry-based science lessons, demonstrations, and experiments. As a follow-up to measure the effectiveness of our institutes, both interviews and surveys were administered in which participants were asked to evaluate how their academic year compared to the quality of science instruction offered prior to their experience in our Science Teaching Institute at Cal Poly Pomona. During our presentation, results of the administered surveys and interviews are shared.

Professional Development for High School Chemistry Teachers Through the Rockefeller University Science Outreach Program

Bonnie L. Kaiser
The Rockefeller University, 1230 York Avenue—Box 53, New York, NY 10065-6399; e-mail: bonnie@rockefeller.edu

Since 1992, the Rockefeller University Science Outreach Program for K-12 teachers and high school students has worked to improve science education through a program of mentored research in the university's 70+ biomedical research laboratories, combined with training in science communication and related student enrichment and teacher professional development activities. Teachers participate for two years and develop action plans for implementing inquiry-based learning in their classrooms. Based on its successful outcomes in general science education and the increased strength of chemistry-related investigations at Rockefeller, in 2007 the program designed a three-year pilot project, Synergy Through Inquiry, to implement and test strategies for improving chemistry education through inquiry-based research and communication training. The components include mentored research in chemistry; a science communication course based on a model paper on evolution at the macromolecular level; a seminar series focusing on "Life as Chemistry and Biological Organization" and featuring chemistry faculty presenting on their research; and Students Modeling a Research Topic (RU-SMART) team collaborations on visualizing chemistry. Synergy Through Inquiry is supported by the Camille & Henry Dreyfus Foundation.

Reaching Rural High School Chemistry Teachers in Florida: Engaging in Scientific Research with Scientists in State Parks, Wildlife Refuges, Estuarine Reserves, and Other Local Resources

Penny J. Gilmer,[1] Amanda Clark, Sarah Sims, Donald Bratton, Joi Walker, Steven Blumsack, Harold Kroto
[1]*Florida State University, Department of Chemistry and Biochemistry, P.O. Box 3064390, Tallahassee, FL 32306-4390; e-mail: gilmer@chem.fsu.edu*

From an in-service program of 79 K-12 teachers, 12 are high school chemistry teachers from rural northwestern and north central Florida. The goal of the two-semester program is to provide opportunities for the teachers to work in collaborative teams with teachers from their rural districts and with scientists that work near their schools. First, in spring 2008, we offered an online graduate class, entitled Nature of Scientific Inquiry, to 118 K-12 teachers, providing discussion boards for students' required weekly posts on relevant readings. The class broadcasts are provided online continuously using a media site at Florida State University at *http://www.geoset.info/sciii/broadcasts.html.*

For those with continued interest, we identified 45 options of research sites spread from western to north central Florida from which the teachers could choose for their graduate class "Scientific Research Experiences." Thirty teams of teachers work together, with two to five teachers per team, for 90 hours of concentrated research experience, supervised by at least one scientist on site. The scientists typically work on environmental issues that take place in state parks, national refuges, estuarine reserves, etc. Teachers reflect in writing on the readings from the scientist and their experiences in the field. At the end of the program we have a poster day in which the teachers present their research in poster format.

We have 12 chemistry high school teachers plus 5 teacher mentors who also are chemists in the program. Their ideas for bringing their learning and experiences to their students include (1) water quality testing of bodies of water local to their schools, (2) "food-for-thought" questions that we used in our online nature-of-scientific-inquiry class, (3) the importance of units of measurement (with practical examples from their research), (4) PBS videos on the *Journey to Planet Earth* series (with broadcasting rights for two years) at *http://www.geoset.info/sciii/JTPE.html*, (5) importance of collaboration and crossover in science, and (6) filtration of methylene blue in different types of soils.

We evaluate the effectiveness of the program using the Views on the Nature of Science questionnaire, before the courses start and at the end of the program. We are particularly interested if having teams with one elementary school, one middle school, and one high school teacher work more effectively with the articulation among the different levels of K-12 schooling. We utilize cultural historical activity theory as our theoretical lens for looking at the coherences and contradictions in the flow of the teachers to their objects and outcomes. We plan to visit some of the teachers in their classrooms early in the upcoming academic year.

A grant from the State of Florida pays the graduate tuition for the teachers and a salary for the summer research. We hire teacher mentors who are K-12 teachers who have done scientific research, and they visit the teams regularly and grade the participants' regular posts. We selected nine participants to write chapters for a monograph on their experiences in the program. We collaboratively work with the Panhandle Area Educational Consortium (PAEC) in Chipley, Fla. Our grant's Webs ite is *http://www.paec-sc-iii.org/index.html*. PAEC is preparing an hour-long video documentary on this program, with visits to the research sites of the nine monograph authors.

Ohio House of Science and Engineering (OHSE), a K-20 Outreach Program

Dr. Susan Olesik

Department of Chemistry, Ohio State University, 100 W. 18th Ave., Columbus, OH 43210, Email: olesik@chemistry.ohio-state.edu

Two programs that are components of the OHSE are highlighted in this poster: Wonders of Our World, W.O.W. and the High School Science Outreach Program of Ohio State University's Nanoscience and Engineering Center.

The Wonders of Our World, W.O.W. is a science outreach program to local K-8 schools. The goals of W.O.W. are to (1) supplement and improve the existing science programs, (2) bring the excitement of science discoveries into the classroom, (3) provide science equipment and content material for teachers, (4) increase community (parents, scientists, and OSU students) involvement in local school activities, and (5) generate a pathway that gives school teachers ready access to scientists at OSU and other local science enclaves. W.O.W. begins its tenth year of operation this fall. This program provides teacher workshops and visits from volunteer scientists throughout the academic year. It is a highly successful

program having served more than 10,000 students and hundreds of teachers to date. Data illustrating its structure and the assessment metrics are provided.

Nanoscience and Technology Experiments to Expand the Capabilities of High School Chemistry, Physics and Biology Teachers, CANPBD Education Committee (S. V. Olesik, D. L. Tomasko, T. Conlisk, and P. R. Kumar): The Ohio State University's Center for Affordable Nanoengineering of Polymeric Biomedical Devices (CANPBD) has established a significant outreach program for in-service high school science teachers. The goals of this effort include (1) introducing high school teachers and students to the excitement of the new discoveries occurring in nanoscience, (2) providing laboratory-based and computer modeling experiments in nanotechnology that are aligned with content standards taught in high school science curricula, (3) illustrating the multidisciplinary nature of most scientific studies, and (4) providing select high school teachers with the opportunity to collaborate with members of the center in developing these experiments. Workshops that allow classroom teachers from across the State of Ohio to work through these experiments and learn more about the center are offered each summer. During the academic year, the members of the center's education committee collect information from participating teachers about how these experiments function in their classrooms. Finally, starting this year an online discussion group has been established to allow facile discussion among the CANPBD scientists and engineers and the high school teachers. This program is beginning its fifth year of evolution. Examples of the experiments developed to date, evaluation metrics, and results are highlighted in this poster.

Promoting Excellence in Science Education Through ACS Outreach Programs

Terri Taylor, Marta Gmurczyk

American Chemical Society, Education Division, 1155 Sixteenth Street, N.W., Washington, DC 20036

With more than 160,000 members, the American Chemical Society is the world's largest scientific society and one of the world's leading sources of authoritative scientific information. A nonprofit organization, chartered by Congress, ACS is at the forefront of the evolving worldwide chemical enterprise and the premier professional home for chemists, chemical engineers, and related professionals around the globe.

The ACS Education Division provides programs, products, and services that promote excellence in science education and community outreach. At the secondary level, these include the High School Chemistry Clubs program, professional development workshops, and a pilot program, Summer Research Fellowships for high school chemistry teachers.

D

Workshop Attendees

Roxie Allen, Association of Chemistry Teachers of Texas (ACT2), Houston

Patricia Baisden, National Nuclear Security Administration/Department of Energy, Washington, DC

L. Anthony Beck, National Institutes of Health, Bethesda, Maryland

Michael Berman, Air Force Office of Scientific Research, Arlington, Virginia

Constance Blaise, University of Pennsylvania, Philadelphia

Paul Bryan, Chevron, Richmond, California

Mark Cardillo, Camille & Henry Dreyfus Foundation, New York

William Carroll, Oxy Chem, Dallas, Texas

Charles Casey, University of Wisconsin, Madison

John Chen, Lehigh University, Bethlehem, Pennsylvania

Katherine Covert, National Science Foundation, Arlington, Virginia

Edward Crowe, Carnegie Corporation, Washington, DC

Hai-Lung Dai, Temple University, Philadelphia, Pennsylvania

Reeny Davison, ASSET Inc., Pittsburgh, Pennsylvania

Jeff Dilks, Department of Energy, Washington, DC

Jay Dubner, Columbia University, New York

Diana Dummitt, University of Illinois at Urbana-Champaign

Caryn Galatis, Thomas A. Edison High School, Alexandria, Virginia

Penny J. Gilmer, Florida State University, Tallahassee

Marta Gmurczyk, American Chemical Society, Washington, DC

Bryce Hach, Hach Scientific Foundation, Fort Collins, Colorado

Myra Halpin, North Carolina School of Sciences and Mathematics, Durham

Kiara Hargrove, Baltimore Polytechnic Institute, Baltimore, Maryland

Alex Harris, Brookhaven National Laboratory, Upton, New York

Sharon Haynie, E. I. du Pont de Nemours and Company, Wilmington, Delaware

Eric Jakobsson, University of Illinois, Urbana-Champaign

Ellen M. Johnson, Wilmington Friends School, Wilmington, Delaware

Bonnie Kaiser, The Rockefeller University, New York

Thomas E. Keller, National Academy of Sciences, Washington, DC

Brian J. Kennedy, Thomas Jefferson High School for Science and Technology, Alexandria, Virginia

Mary Kirchoff, American Chemical Society, Washington, DC

Michael Klein, University of Pennsylvania, Philadelphia

Jeff Krause, Department of Energy, Washington, DC

Sandra Laursen, University of Colorado, Boulder

Marshall Lih, National Science Foundation, Arlington, Virginia

Paul McKenzie, Centocor R&D, Radnor, Pennsylvania

Jin Montclare, State University of New York, Brooklyn

Sergey Nizkorodov, University of California, Irvine

Susan V. Olesik, Ohio State University, Columbus

Gilbert Pacey, Miami University, Oxford, Ohio

Michael Page, California State Polytechnic University, Pomona

Joan T. Prival, National Science Foundation, Arlington, Virginia

Marquita Qualls, Entropia, West Chester, Pennsylvania

Michael Rogers, National Institutes of Health, Bethesda, Maryland

Sophie Rovner, Chemical and Engineering News, Washington, DC

Kim Schaaf, PPG Industries, Allison Park, Pennsylvania

Hratch Semerjian, The Council for Chemical Research, Washington, DC

Loraine Snead, Wilmington Friends School, Wilmington, Delaware

Patricia Soochan, HHMI, Chevy Chase, Maryland

Kaye Storm, Stanford University, Stanford, California

Kathryn Sullivan, Battelle Center for Mathematics & Science Education Policy, Columbus, Ohio

Robert Tai, University of Virginia, Charlottesville

Irwin Talesnick, Queens University, Ontario, Canada

Terri Taylor, American Chemical Society, Washington, DC

Gerry Wheeler, National Science Teachers Association, Washington, DC

Ken White, BNL Office of Educational Opportunity, Upton, New York

Stacie Williams, University of Dayton, Dayton, Ohio

Edee Wiziecki, University of Illinois at Urbana-Champaign

E

Origin of and Information on the Chemical Sciences Roundtable

In April 1994 the American Chemical Society (ACS) held an Interactive Presidential Colloquium entitled "Shaping the Future: The Chemical Research Environment in the Next Century."[1] The report from this colloquium identified several objectives, including the need to ensure communication on key issues among government, industry, and university representatives. The rapidly changing environment in the United States for science and technology has created a number of stresses on the chemical enterprise. The stresses are particularly important with regard to the chemical industry—a major segment of U.S. industry that makes a strong, positive contribution to the U.S. balance of trade and provides major employment opportunities for a technical workforce. A neutral and credible forum for communication among all segments of the enterprise could enhance the future well-being of chemical science and technology.

After the report was issued, a formal request for such a roundtable activity was transmitted to Dr. Bruce M. Alberts, chairman of the National Research Council (NRC), by the Federal Interagency Chemistry Representatives, an informal organization of representatives from the various federal agencies that support chemical research. As part of the NRC, the Board on Chemical Sciences and Technology (BCST) can provide an intellectual focus on issues and fundamentals of science and technology across the broad fields of chemistry and chemical engineering. In the winter of 1996, Dr. Alberts asked BCST to establish the Chemical Sciences Roundtable to provide a mechanism for initiating and maintaining the dialogue envisioned in the ACS report.

The mission of the Chemical Sciences Roundtable is to provide a science-oriented, apolitical forum to enhance understanding of the critical issues in chemical science and technology that affect the government, industrial, and academic sectors. To support this mission the Chemical Sciences Roundtable will do the following:

- Identify topics of importance to the chemical science and technology community by holding periodic discussions and presentations, and gathering input from the broadest possible set of constituencies involved in chemical science and technology.
- Organize workshops and symposia and publish reports on topics important to the continuing health and advancement of chemical science and technology.
- Disseminate information and knowledge gained in the workshops and reports to the chemical science and technology community through discussions with, presentations to, and engagement of other forums and organizations.
- Bring topics deserving further in-depth study to the attention of the NRC's Board on Chemical Sciences and Technology. The roundtable itself will not attempt to resolve the issues and problems that it identifies—it will make no recommendations, nor provide any specific guidance. Rather, the goal of the roundtable is to ensure a full and meaningful discussion of the identified topics so that participants in the workshops and the community as a whole can determine the best courses of action.

---

[1]American Chemical Society. 1994. *Shaping the Future: The Chemical Research Environment in the Next Century.* American Chemical Society Report from the Interactive Presidential Colloquium, April 7-9, Washington, DC.

58